Claude Louis-Charles PhD

Serious Managers' Guide to Responsible AI

A Detailed Playbook for Managing AI Risk, Fairness, and Compliance

Claude Louis-Charles PhD

Serious Managers Guide to Responsible AI

The moral rights of the author have been asserted

Copyrights 2026

Cybersoft Publishing LLC
Fort Washington, MD 20744

DrClaude.net

First Edition March 2026

Claude Louis-Charles PhD

Table of Contents

1   New Mgmt. Mandates: AI with Accountability 1

1.1   Redefining Responsible AI for IT leaders 3

1.2   Governance as an enabler, not a blocker 5

1.3   The Manager's new responsibilities 6

1.4   Business case for responsible modernization .. 8

1.5   The Monday Morning Test 10

2   Why Ethics Is a Business Imperative 13

2.1   Why this matters for AI modernization 13

2.2   Demystifying AI ethics for IT managers 14

2.3   The cost of neglecting ethics 17

2.4   Ethics as a driver of competitive advantage ... 18

2.5   Translating values into concrete AI principles 19

2.6   The IT manager's ethics checklist 20

2.7   Empowering teams to flag ethical concerns... 25

3   Safety First: Building Trust Through Protection 31

3.1   What AI safety really means for leaders 32

3.2   From robustness to real-world risk 34

3.3   The Safety by Design framework 36

3.4   Identifying concrete harm scenarios 37

3.5   Misuse deterrence: anticipating adversaries .. 38

3.6   Org. resilience: planning for AI incidents 39

3.7   Safety roles and responsibilities 40

3.8   The Manager's Safety Playbook 45

4   Fairness at Scale 50

4.1   Why fairness is a business and mission issue 50

4.2   Fairness is not a one-time configuration 51

4.3   Understanding common sources of bias 53

4.4   Manager's fairness checklist (project level) ... 56

4.5   Designing fairness checks into workflows 57

4.6   Bias Audit Checklist 60

4.7   Gov. mechanisms that make fairness stick 66

4.8   Manager's Fairness Playbook (90-day plan) ... 68

4.9   The Monday Morning Test 71

5   Transparency: Seeing Inside the Black Box 73

5.1   Transparency the foundation accountability .. 73

5.2   The building blocks of AI transparency 76

5.3   Model cards: the "spec sheet" for AI models . 77

5.4   AI fact sheets: broadening beyond the model 81

5.5   Manager checklist: transparency and logging 86

5.6   Comm. strategies: how you talk about AI 89

5.7   AI-Supported Decision Notice 91

5.8   Executive AI Transparency Brief 92

5.9   Reality Chk:avoiding "black-box comfort." 94

6   The Accountability Chain 98

   6.1   Accountability - backbone of modernization . 98

   6.2   Accountability across the AI lifecycle 100

   6.3   Mapping the Accountability Chain: 101

   6.4   Rights: who decides what, and when 104

   6.5   Appr. checkpoints: building control flow 105

   6.6   Escal. paths: When something goes wrong .. 108

   6.7   RACI for AI accountability 111

   6.8   Mgr. checklist: accountability chain real? 112

   6.9   Vendor vs internal accountability 114

7   Policy and Compliance in Practice 121

   7.1   Compliance:paperwork/prevention 123

   7.2   Evolving AI policy landscape 124

   7.3   Translating external rules into policies 125

   7.4   The manager's AI policy checklist 126

   7.5   The EU AI Act risk-based approach 128

   7.6   Building compliance with workflows 132

   7.7   AI project intake & policy trigger form 133

   7.8   Compliance, what you will be asked to show 135

   7.9   AI Reality Check: avoiding "policy theater." . 142

8   Managing AI Risk: Frameworks That Work 145

   8.1   Risk frameworks - heart of governance 145

8.2   The core idea: value, feasibility, and impact. 147

8.3   Introducing ISO 42001: AI mgmt systems..... 149

8.4   Making frameworks usable 150

8.5   Domain adaptations: 156

8.6   Portfolio-level risk: seeing across systems .. 160

8.7   AI Risk Portfolio Summary – Template 161

8.8   Mgrs AI Risk Playbook (90-day plan) 163

8.9   Avoiding "framework theater." 164

9   The Governance Operating Model 167

9.1   Why AI governance operating model matters167

9.2   Governance as workflow, not bureaucracy .. 169

9.3   "AI Governance Oper. Model" framework 170

9.4   Flow: the AI governance lifecycle 174

9.5   AI Governance Workflow – Project Checklist 175

9.6   Controls: calibrating governance to culture . 176

9.7   Signals: metrics that show governance........ 179

9.8   Governance Oper. Model Health Checklist .. 179

9.9   Right-sizing govern. matching model to org.. 181

9.10   AI Govern. Oper. Model – Design Canvas 182

9.11   The Mgr's Govern. Playbook (90-day plan) ... 184

9.12   Governance that actually changes behavior 185

9.13   The Monday Morning Test........................... 186

10   Embedding Responsibility by Design 189

10.1   Responsibility must live inside delivery........ 189

10.2   Responsibility by design: the core concept .. 191

10.3   The Responsibility by Design lifecycle 192

10.4   Scope & intake responsibility checklist........ 194

10.5   Design & procurement responsibility 195

10.6   Responsible AI Design Brief – Template 196

10.7   Responsible AI Vendor Questionnaire 205

10.8   Responsibility into governing. processes 206

10.9   Avoiding "responsibility theater." 209

11   Data Integrity and Stewardship 212

11.1   Data integrity the foundation of Respons. AI 212

11.2   What "data governance" means in the AI era 214

11.3   Knowing where your AI's data comes from... 215

11.4   Labeling quality and annotation integrity 218

11.5   Data trustworthiness: defining quality for AI. 220

11.6   Introducing the Data Stewardship Board 222

11.7   Data roles: owners, stewards, custodians ... 225

11.8   Data trustworthiness checklist 226

12   Culture, Training, and Human Oversight.............. 233

12.1   Why culture matters as much as controls.... 233

12.2   Elements of a "responsibility culture." 235

12.3 Training: making govern. "how we work."..... 236

12.4 Responsible AI training checklist................. 237

12.5 Decision checklist essentials 238

12.6 Oversight: what it is and what it is not.......... 239

12.7 Designing human oversight into workflows .. 240

12.8 Human Oversight Specification – Template . 242

12.9 Psychological safety signals....................... 244

12.10 The AI Reality: culture is the multiplier 248

13 Measuring and Reporting AI Impact..................... 251

13.1 Why governance needs metrics.................. 251

13.2 Designing responsible AI KPIs: principles..... 253

13.3 Ethics and fairness KPIs 254

13.4 Ethics & fairness KPI checklist (per system) . 254

13.5 Compliance and governance KPIs 256

13.6 Value and adoption KPIs 257

13.7 Building AI impact scorecard (system level) 258

13.8 Portfolio-level AI impact reporting 259

13.9 External reporting: regulators and the public 262

13.10 Turning scrutiny into a trust asset 263

14 The Responsible AI Roadmap 269

14.1 You need a roadmap, not more principles ... 269

14.2 The four-phase Responsible AI roadmap 270

14.3   Assess phase checklist 272

14.4   Responsible AI baseline assessment summ 272

14.5   Template: Pilot project Responsible AI plan . 275

14.6   Conn. the roadmap to business outcomes .. 280

14.7   Sequencing: which levers to pull first 282

14.8   Mge's Integrated Roadmap Checklist 283

14.9   Your first 90 days from this book 285

14.10   The final Monday Morning Test 286

15   Responsible AI and Governance 287

15.1   Contextualizing AI Governance 287

15.2   AI Impacts: Positives, Perils, and Risks 288

15.3   Responsible AI Principles 300

15.4   Principles and Frameworks AI Governance . 306

15.5   Principles of AI Governance (Whys) 308

15.6   Frameworks for AI Governance (Hows) 312

15.7   Interplay Between Principles and Frmwks ... 318

15.8   Conclusion 321

16   The Legal Framework 322

17   Content References 334

1 New Mgmt. Mandates: AI with Accountability

Opening vignette

It's Monday morning in a mid-size government agency. The IT modernization lead is presenting a new AI-enabled case management system to the leadership team. The demo wows the room—faster routing, fewer manual tasks, smarter insights. But then the CFO asks a question that stalls the momentum: "What happens if the model's recommendation is wrong? Who's accountable for that?" The room goes silent. Modernization leads to a sudden realization that this isn't just a technological upgrade, it's a shift in management and governance.

Why this matters now

AI modernization changes everything about how IT managers lead. It blurs the lines between human and machine decision-making, between technical performance and ethical responsibility. For decades, IT projects were measured by uptime, cost efficiency, and user adoption. Now, executives are asking:

- How is this model making its decisions?
- Who approves of its deployment?
- What happens if bias or harm occurs?

This chapter positions **responsible AI and governance** not as compliance tasks, but as **core management capabilities** required to scale AI safely, sustainably, and credibly.

The AI Accountability Shift

From technology projects to management discipline

In traditional IT modernization, managers lead system upgrades and process automation. With AI, that mandate expands you're shaping how organizational decisions are made.

AI systems learn, adapt, and sometimes behave unpredictably. Managing them requires continuous oversight, not one-time implementation.

In this new management era:

- **Risk** moves from technical failure to decision failure.

- **Accountability** shifts from code ownership to outcome ownership.
- **Governance** becomes the active control system of your AI operation.

From Project Management to AI Governance Cycle

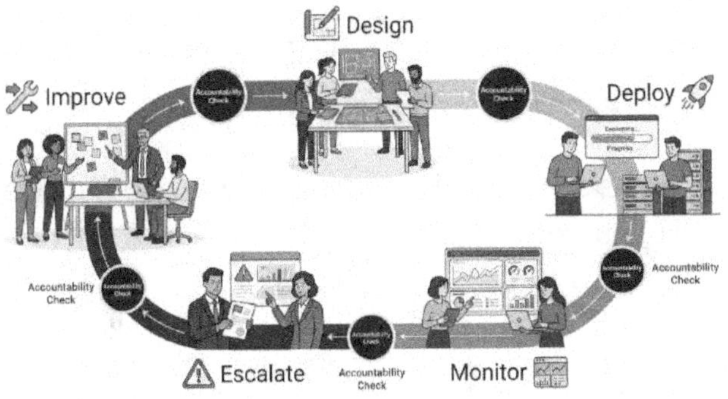

1.1 Redefining Responsible AI for IT leaders

Responsible AI isn't abstract ethics—it's operational management of risk, fairness, and compliance in every workflow. For IT managers, it boils down to three core tasks:

1. **Set clear guardrails.** Define what AI can and cannot do within your modernization program.
2. **Operationalize oversight.** Embed monitoring, human review, and auditability into every system.

3. **Report with transparency.** Make decisions traceable, explainable, and justifiable to non-technical leaders.

These actions transform responsible AI from a concept into a management system.

The convergence of three pressures

AI modernization forces IT managers to respond to three converging forces:

- **Regulatory expectations.** Governments are issuing new AI risk mandates (like transparency reporting, documentation, and model explainability). Your modernization plan must align with these emerging standards.
- **Public trust.** AI systems interacting with customers or citizens are judged not just on performance but on fairness and safety. Perception risk is now operational risk.
- **Organizational risk appetite.** C-suites and boards want innovation—but not at the expense of credibility or compliance. IT managers must translate these expectations into practical guardrails.

(Insert Diagram 3: "The Pressure Triangle" – blue triangular diagram labeled **Regulation**, **Trust**, and **Risk Appetite**, with "Responsible AI Governance" at the center.)

1.2 Governance as an enabler, not a blocker

Too often, governance is seen as a form of bureaucracy. In practice, strong AI governance accelerates modernization by creating clarity: who decides, who monitors, and what happens when things go wrong.

For IT managers, governance provides four operational benefits:

- **Clarity of roles.** Everyone knows who owns which decision.
- **Reduced rework.** Early risk identification avoids downstream failures.
- **Faster approvals.** Clear criteria speed up deployment.
- **Trustworthy innovation.** Leadership feels confident green-lighting AI at scale.

Governance as Flow, Not Friction

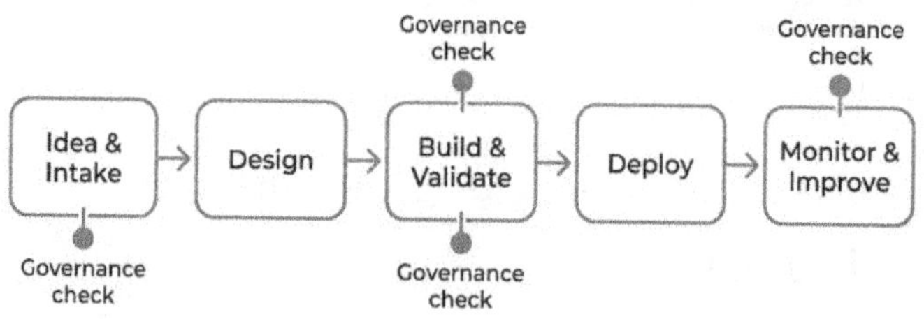

1.3 The Manager's new responsibilities

In traditional IT roles, managers control infrastructure. In AI modernization, they must control **outcomes**. That means:

- Translating ethical principles into technical requirements.
- Setting escalation paths when AI outcomes deviate.
- Balancing innovation against legal and reputational risk.
- Building cross-functional governance with legal, data, and business teams.

Expanding Manager Responsibilities

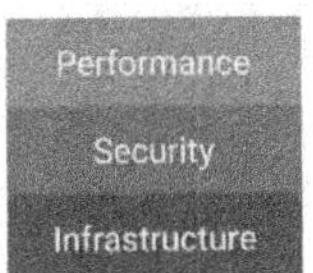

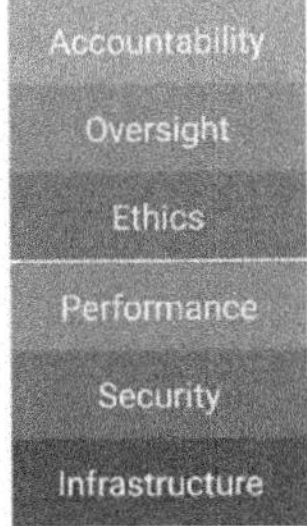

Building the AI accountability chain

Accountability must be visible, traceable, and enforceable. IT managers should establish a simple **AI Accountability Chain**, linking every stage of the system lifecycle:

- **Data Owners** ensure quality and representativeness.
- **Model Developers** document design and limitations.
- **Solution Owners** approve deployment conditions.
- **Operators** monitor performance and flag anomalies.
- **Executives** own the outcomes and external reporting.

AI Accountability Chain

Clear links prevent responsibility drift.

1.4 Business case for responsible modernization

Responsible AI is not just ethical—it is strategic. Organizations that manage AI risk well:

- Deploy faster because they preempt regulatory blockers.
- Spend less on rework and incident mitigation.
- Earn public and internal trust faster.
- Attract talent proud to build accountable systems.

Real ROI comes from predictability. Responsible processes make modernization sustainable, reducing both technical debt and reputational exposure.

Turning accountability into action

How can IT managers implement accountability starting this quarter?

1. **Assess current governance maturity.** Identify gaps in documentation, monitoring, and decision rights.
2. **Form an AI review board.** Include representatives from legal, ethics, security, and user operations.
3. **Make traceability mandatory.** Every system must produce an audit trail of how its decisions were made.
4. **Pilot "accountability dashboards."** Display fairness metrics, audit frequency, and issue logs.
5. **Report to leadership.** Redefine project success beyond delivery to include safety and trust metrics.

Case example: An AI-powered customer support system

The federal IT department deployed an AI chatbot to triage citizen inquiries. Early tests showed faster responses but emerging fairness issues: non-native English speakers received lower service ratings. Instead of shelving the system, the IT manager established a fairness audit team, retrained the model, and built real-time quality dashboards. Within a quarter, bias scores dropped 30%, and satisfaction improved 15%.

This story illustrates responsible governance not as restriction but continuous improvement.

1.5 The Monday Morning Test

On Monday morning, the IT manager responsible should be able to answer five questions:

1. Which AI systems under my scope make consequential decisions?
2. For each one, who owns outcomes if it fails?
3. How does my team detect and escalate AI issues?
4. Which governance checkpoints are automated versus manual?
5. How do we demonstrate accountability to leadership and regulators?

If you cannot answer all five yet, this book will help you build those capabilities step by step.

The Monday Morning Checklist

1	Which AI systems under my scope make consequential decisions?
2	For each one, who owns outcomes if it fails?
3	How does my team detect and escalate AI issues?
4	Which governance checkpoints are automated versus manual?
5	How do we demonstrate accountability to leadership and regulators?

1.5.1 Chapter takeaway

AI modernization is no longer only about driving efficiency—it is about earning and maintaining trust. For IT managers, accountability is not optional; it is the foundation of their relevance as leaders in the AI era. Responsible AI and governance do not slow modernization—they make it stick.

Create a modern-looking image that describes the above, use a blue palette, and leverage imagery. Add people to the image. Create a modern-looking image that describes the above, use a blue palette, and leverage imagery. Add people to the image.

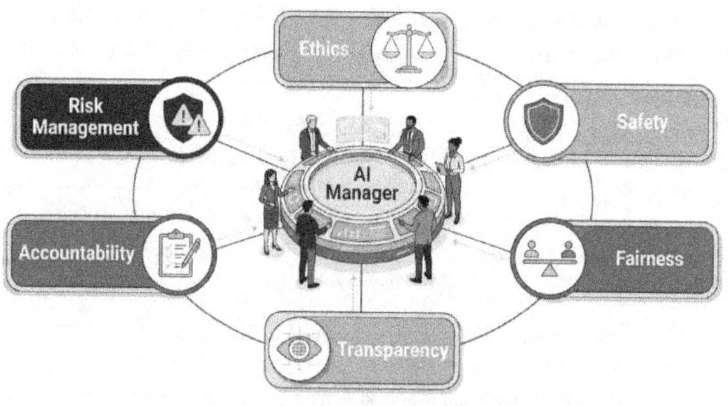

2 Why Ethics Is a Business Imperative

Opening vignette

You are leading an AI modernization program for a major enterprise. Your team just rolled out an AI-driven resume screening tool to speed up hiring for critical technical roles. Within weeks, recruiters are thrilled: the timetable has dropped by 60 percent. Then HR brings you an uncomfortable report: applicants from certain universities and zip codes are barely making it through the screening process. A few candidates post their stories on social media. A journalist emails your communications team asking for comments. The CIO calls you: "Walk me through how this system makes decisions—and whether we have an ethics problem."
In that moment, it is clear: ethics is not an abstract concept. It is directly tied to brand, talent, regulatory exposure, and your credibility as an IT leader.

2.1 Why this matters for AI modernization

For IT managers, AI ethics is no longer a "nice to have" handled by philosophy teams or occasional training slides. It is a core part of designing, deploying, and defending AI systems. When AI touches hiring, credit, benefits eligibility, case routing, customer service, or clinical support, ethical missteps create real harm: to people, to your organization's reputation, and to your modernization roadmap.

Ethical failures derail projects, trigger urgent remediation work, and invite scrutiny at exactly the moment you need support to scale AI. Conversely, when you embed ethics into your modernization program from the start, you reduce the odds of failure, build trust with leadership, and create a durable license to operate.

Ethics as a Risk and Value Lever

2.2 Demystifying AI ethics for IT managers

Many managers hear "AI ethics" and think of complex philosophical debates, far removed from sprint planning and architecture decisions. For modernization leaders, the practical definition is simpler: **AI ethics is the discipline of making sure your AI systems align with your organization's values and obligations in real operational workflows.**

In practice, that means asking structured questions before and after deployment:

- Who could be harmed by this system, even unintentionally?
- Who benefits—and is that benefit fairly distributed?
- Are we transparent enough that we would be comfortable explaining this to a regulator, a newspaper, or our own employees?
- Do people impacted by AI decisions have ways to question or appeal them?

Ethics becomes a lens you apply continuously, not a single checkbox at the end of the project.

From Abstract Ethics to Practical Questions

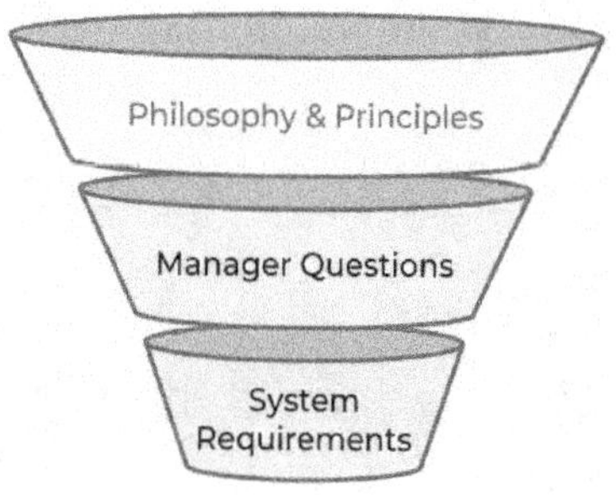

How ethics shows up in day-to-day decisions

You already make ethical calls in your IT role, even if you do not label them that way. With AI, those calls become more frequent and more consequential. Examples include:

- Deciding whether to use historic data that reflects legacy bias.
- Choosing whether a model can auto-approve a loan or only make a recommendation.
- Setting confidence thresholds for automated decisions versus human review.
- Determining how much explanation the system must give to users.

Each of these choices has ethical implications: who is included or excluded, who bears the risk, and how power is distributed between the organization and the people impacted by AI. Ethics is not a separate track; it is embedded in architecture, configuration, and workflow design.

2.3 The cost of neglecting ethics

Neglecting ethics is expensive in ways that are both visible and hidden. In visible terms, organizations face public backlash, formal complaints, regulatory investigations, and loss of customers or citizens' trust. Hidden costs include delayed projects, extra change-management work, and staff demoralization when teams feel they are shipping systems that conflict with their values.

For IT managers, ethical failures often trigger crisis mode: emergency patching of models, sudden feature rollbacks, unplanned audit work, and intense executive scrutiny. This noise pulls your team away from forward-looking modernization and traps you in reactive firefighting. The lesson: ethics is cheaper upstream.

Upstream vs Downstream Cost of Ethics

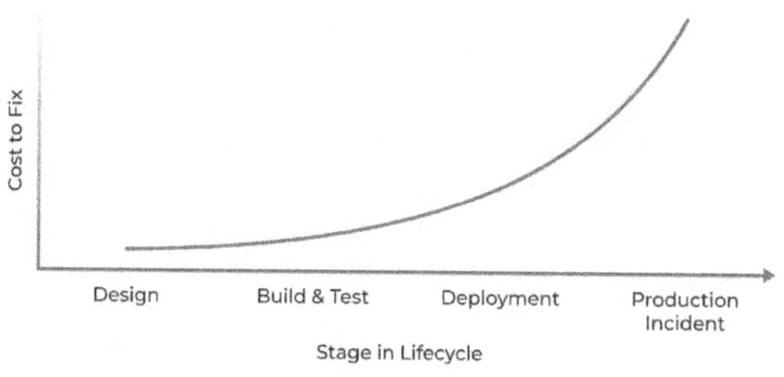

2.4 Ethics as a driver of competitive advantage

Ethics is not just about avoiding harm; it is about creating advantages. When your AI systems are seen as fair, transparent, and respectful, several benefits follow:

- Customers are more willing to adopt AI-driven services.
- Partners and regulators treat your organization as a responsible actor, easing approvals.
- Talent (especially technical and early-career) is more likely to join and stay.
- Internal stakeholders trust your modernization program and support its expansion.

Ethical leadership becomes part of your brand. Organizations known for careful, accountable AI gain a reputational moat that is difficult for competitors to copy quickly. For an IT manager, this is a strategic lever—ethics becomes part of your modernization story rather than an afterthought.

2.5 Translating values into concrete AI principles

Most organizations already have value statements: respect, integrity, fairness, inclusion, and safety. Your job as an IT modernization leader is to translate those broad values into AI-specific principles that can guide design and implementation. For example:

- "We treat people fairly" becomes "We minimize and monitor disparate impact in our models."

- "We are transparent" becomes "We provide clear explanations and accessible documentation of AI-supported decisions."
- "We respect individuals" becomes "We give people meaningful ways to contest or appeal AI-influenced decisions."

This translation step turns corporate values into actionable items for engineers, product owners, and data teams.

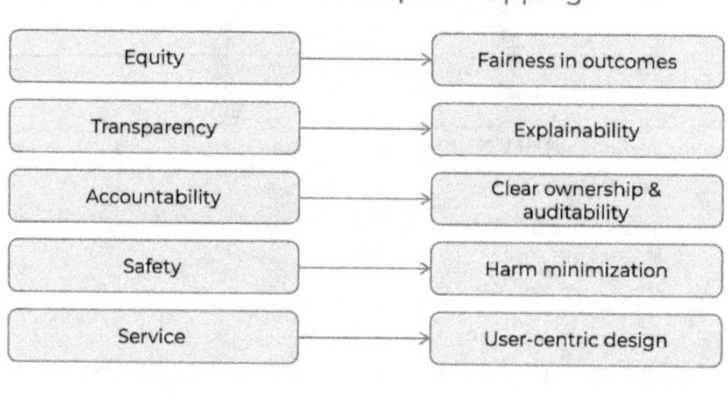

Values to AI Principles Mapping

2.6 The IT manager's ethics checklist

To operationalize ethics, you need simple, repeatable tools—starting with a checklist you can apply across AI projects. A practical ethics checklist for IT managers may include questions like:

- Purpose: Is this AI system solving a legitimate problem aligned with our mission?

- Necessity: Do we need AI for this, or is there a simpler, less risky solution?
- Impact: Who is affected, and have we considered vulnerable groups?
- Data: Where does the data come from, and does it encode historic bias?
- Oversight: What decisions are fully automated versus requiring human oversight?
- Recourse: How can people raise concerns or challenge outcomes?
- Documentation: Can we explain how it works and why we built it this way?

You can incorporate this checklist into intake forms, architecture reviews, and go-live approvals so it becomes part of standard operating procedure, not a special-case process.

Embedding ethics into your governance structure

Ethics cannot rely on one passionate individual. It must be anchored in your governance structure. As an IT manager, you can champion the creation or strengthening of:

- An AI review board or working group that evaluates high-risk use cases.
- Clear criteria for "ethically sensitive" projects needing additional scrutiny.
- Standardized documentation templates that capture ethical considerations.
- Escalation paths when issues are discovered post-deployment.

By building ethics into your governance forums, you give teams a place to raise concerns, get guidance, and resolve dilemmas before they become incidents.

Real-world pattern: biased recruitment algorithms

Consider a common AI modernization use case: automating parts of the recruitment process. A model is trained on historic hiring data to predict which resumes are likely to result in successful hires. If the historical data reflects an underrepresentation of certain groups, the model may learn to devalue resumes from those groups, even if it never directly sees protected attributes.

For the IT manager, the ethical questions include:

- Should we use this historical data as-is?
- Do we need to add fairness constraints or re-sampling techniques?
- Should the AI only rank candidates, while humans make final decisions?

- How do we monitor outcomes to ensure patterns do not drift over time?

By asking these questions early, you avoid deploying a system that quietly amplifies past inequities and exposes your organization to risk.

Biased Recruitment Risk Flow

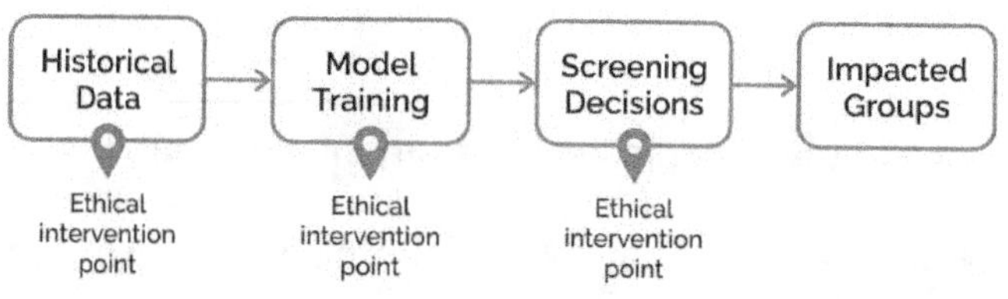

Real-world pattern: opaque credit or eligibility scoring

In many modernization programs, AI is used to score risk — creditworthiness, benefits eligibility, fraud likelihood, or case prioritization—thereby influencing who gets access to resources and who faces additional scrutiny. Opaque scoring models create ethical pressure when people affected cannot understand why they received a decision, or when patterns of exclusion emerge along demographic lines.

As an IT leader, you must determine:

- The level of explainability required for decisions.
- What information users should see about how scores are calculated.
- How your team will test for unintended systemic bias.
- Under what conditions are AI scores advisory versus determinative?

This is not only about protecting your organization; it is about honoring the dignity of people who are more than just entries in a dataset.

Transparency vs Risk Matrix

2.7 Empowering teams to flag ethical concerns

Ethics is not a one-person job. Your engineers, analysts, and product owners see issues at the ground level—if they feel safe to speak up. Part of your role is building a culture where raising ethical concerns is rewarded, not punished. This includes:

- Explicitly inviting ethical questions in design reviews and retrospectives.
- Protecting team members who surface uncomfortable risks.
- Celebrating when teams decide to change direction based on ethical considerations.
- Providing clear channels (formal and informal) to escalate concerns.

A culture of ethical candor is a competitive asset. It helps you detect and address issues before they reach the outside world.

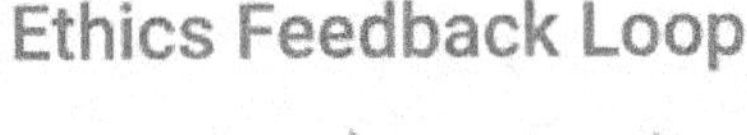

Ethics Feedback Loop

The Manager's Playbook: Putting ethics into motion

To make ethics truly operational in your AI modernization program, you can take a structured approach. A simple playbook for the next 90 days might include:

1. **Clarify principles.** Work with leadership to define a short set of AI ethics principles aligned to organizational values.
2. **Embed questions.** Integrate ethical questions into project intake, architecture reviews, and deployment gates.

3. **Train your teams.** Run short, practical workshops for engineers and product owners on how to use the ethics checklist.
4. **Select pilots.** Choose one or two active AI projects to pilot enhanced ethical review and document lessons learned.
5. **Create visibility.** Report to leadership not only on delivery milestones but also on the identification and mitigation of ethical risks.

This playbook ensures ethics is part of "how we work," not "what we talk about when there is a scandal."

90-Day Ethics Playbook Roadmap

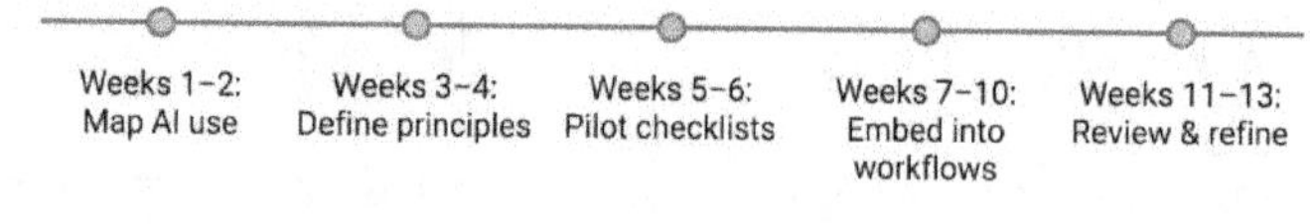

The AI Reality Check: Avoiding ethics theater

There is a real risk of "ethics theater" where organizations publish principles, host panels, and create glossy decks but do not change how they design and deploy systems. As an IT manager, you guard against this by insisting that ethical commitments show up in:

- Requirements documents and user stories.
- Acceptance criteria and test plans.
- Monitoring dashboards and incident runbooks.

If you cannot point to a concrete artifact or process where ethics is implemented, you likely have optics rather than operationalization. Ethics must live where work happens—in tickets, in code, in workflows—not just on posters.

Ethics Theater vs Ethics in Practice

The Monday Morning Test

By the end of this chapter, you should be able to sit down on Monday morning and answer these questions for your AI modernization portfolio:

1. For each high-impact AI system, can I articulate the ethical risks we considered before deployment?
2. Do my teams know which ethical questions they are expected to ask during design and review?

3. Can I show at least one concrete artifact (template, checklist, guideline) where ethics is embedded in our workflow?
4. Do we have a defined path to escalate ethical concerns beyond the project team?
5. Would I feel comfortable explaining our AI systems' purpose, data, and oversight model to a skeptical regulator or journalist?

If the answer to some of these is "not yet," you now have a clear direction. Ethics is not about having all the answers immediately; it is about committing to responsible questions and building processes that keep those questions alive throughout your AI modernization journey.

Ethics Readiness Gauge

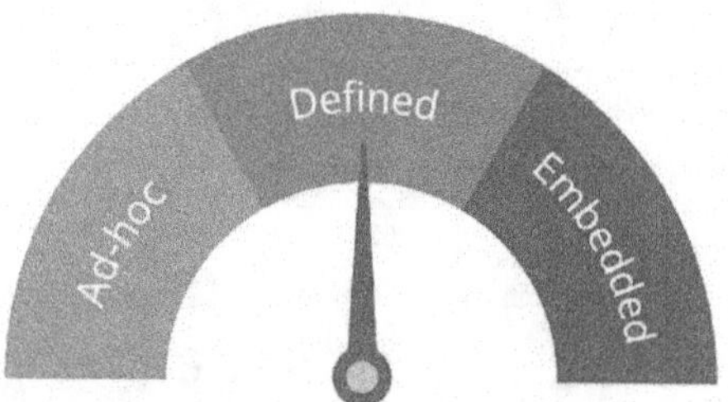

Informal self-assessment tool for managers.

3 Safety First: Building Trust Through Protection

Opening vignette

You lead an AI modernization program for a large enterprise. A new AI assistant has been rolled out to help frontline staff answer customer questions and process simple requests. Early results look great: fewer calls, shorter handle times, happier staff. Then a security operations analyst flags an incident: the assistant was persuaded by a clever prompt to reveal internal troubleshooting steps and draft language that, if sent externally, could mislead customers about policy. Marketing is alarmed, Legal wants details, and the CISO calls you: "Walk me through how this system is controlled—what are the safety guardrails?" In that moment, you realize AI safety is not just about "does it work?" It is about "what happens when things go wrong, and how do we prevent harm before it happens?"

Why safety is now a frontline management issue

In traditional IT projects, safety risks were primarily related to outages, data breaches, and performance failures. With AI, the risk surface expands dramatically: systems can generate misleading content, make autonomous decisions, be manipulated by adversarial prompts, and be repurposed in ways you did not intend. As an IT manager leading AI modernization, you are not just accountable for keeping systems up and running—you are accountable for preventing harm to people, operations, and the organization's reputation.

AI safety has moved from a specialized technical concern to a core management responsibility. Your leadership credibility depends on your ability to identify risks, prioritize mitigations, and design systems that fail safely rather than catastrophically.

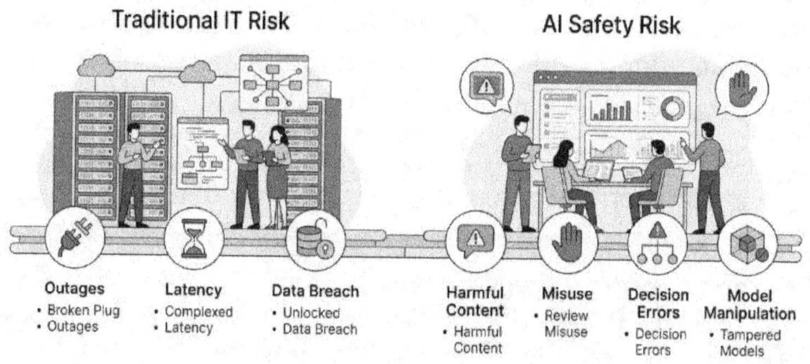

3.1 What AI safety really means for modernization leaders

AI safety often sounds abstract, but for modernization leaders, it breaks down into three concrete domains:

- **Harm prevention:** Avoiding direct harm to users, customers, employees, and the public (for example, incorrect medical guidance, discriminatory decisions, or harmful content).
- **Misuse deterrence:** Reducing the likelihood that internal or external actors can weaponize your AI systems (for example, generating scams, evading controls, or exfiltrating sensitive information).
- **Organizational resilience:** Ensuring your organization can detect, contain, and recover from AI-related incidents without prolonged disruption.

AI safety is not a static property; it is a set of practices you embed across the lifecycle of every AI system.

Three Pillars of AI Safety

3.2 From robustness to real-world risk

Technical robustness—how well a model performs against test data—matters. However, safety goes beyond test metrics to real-world consequences. A model can be highly accurate on average and still be unsafe if it fails in ways that cause concentrated harm to specific groups or in high-stakes scenarios.

As an IT manager, you must distinguish between:

- **Performance risk:** The model is inaccurate or unstable.
- **Safety risk:** The model's behavior in context can cause unacceptable harm, even if technically "working as designed."

For example, an AI classifier that occasionally mislabels a marketing image poses a low safety risk. In contrast, an AI system that occasionally misroutes emergency service calls poses a very high safety risk, even at similar error rates.

Performance vs Safety Risk Grid

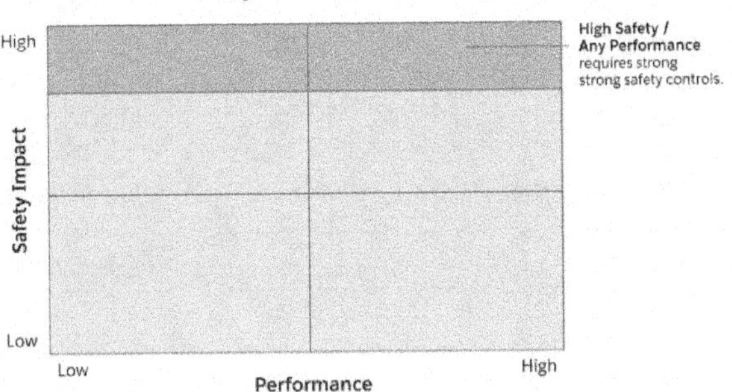

Safety by Design mindset

Most organizations are used to "fix it later" patterns: ship first, patch safety and security issues in response to incidents. With AI, this approach is too slow and too risky. **Safety by Design** means you treat safety not as a bolt-on, but as a core design principle from the first conversation about an AI use case.

For modernization leaders, Safety by Design involves:

- Challenging whether an AI system should be built at all for certain use cases.
- Defining harm scenarios and abuse cases up front.
- Designing workflows where humans stay in the loop for high-risk decisions.
- Building monitoring and control surfaces before deployment—not afterward.

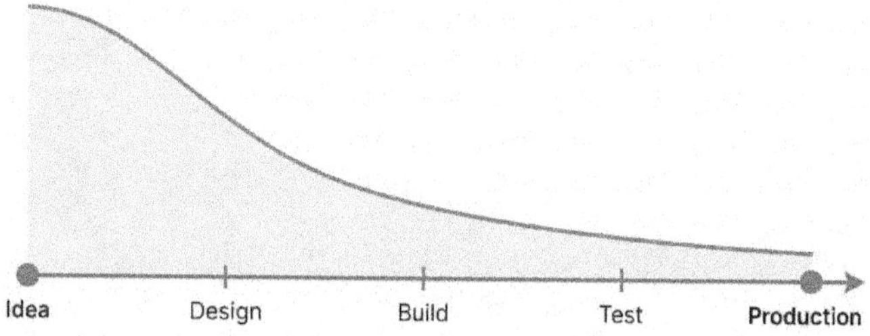

3.3 The Safety by Design framework

To make Safety by Design actionable, you can use a simple three-stage framework across the AI lifecycle: **Design, Deploy, Defend**.

1. **Design:** Identify harms, define safeguards, constrain capabilities, and select appropriate data and models.
2. **Deploy:** Control access, implement guardrails (like content filters and policy constraints), and validate behavior in realistic conditions.
3. **Defend:** Continuously monitor for misuse and failures, respond to incidents, and update systems based on real-world feedback.

This framework gives you a clear structure for decisions, documentation, and accountability.

Safety by Design: Design → Deploy → Defend

3.4 Identifying concrete harm scenarios

To manage safety effectively, you must move beyond abstract fears to concrete harm scenarios. A helpful exercise for each AI system is to list:

- **Intended use:** What is this system supposed to do, for whom, and under what conditions?
- **Misuse scenarios:** How could a bad actor intentionally misuse this system?
- **Failure scenarios:** How could the system accidentally cause harm through normal use?
- **Escalation scenarios:** When should the system hand off to a human or shut down?

For example, a generative AI assistant in a call center might have misuse scenarios, such as being prompted to generate fraudulent instructions, and failure scenarios, such as confidently giving outdated policy guidance.

3.5 Misuse deterrence: anticipating adversaries

AI systems are attractive targets. Prompt injections, model extraction, data poisoning, and abuse of generative capabilities are no longer theoretical. As an IT manager, you need to treat misuse deterrence as part of your safety program, not solely a security team concern.

Practical steps include:

- Strictly controlling which models can connect to internal systems and data sources.
- Implementing input and output filters for harmful content, personally identifiable information, and sensitive topics.
- Rate-limiting high-risk operations and monitoring for unusual usage patterns.
- Separating internal-only and external-facing capabilities.

Your goal is not to eliminate all misuse (impossible), but to raise the cost and lower the impact of abuse.

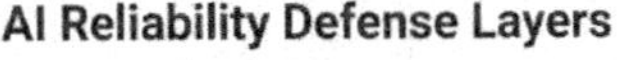

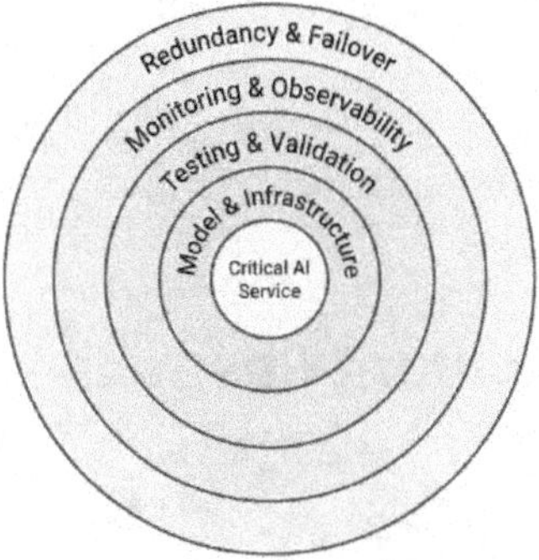

3.6 Organizational resilience: planning for AI incidents

No matter how careful you are, incidents will happen. Safety means assuming failure and designing for resilience. That requires treating AI incidents with the same rigor as security incidents or outages.

Key elements of AI resilience include:

- **Detection:** How will you know when an AI system is behaving dangerously or being abused?
- **Containment:** How quickly can you throttle, isolate, or temporarily turn off problematic functionality?
- **Communication:** Who needs to know, and how will you communicate internally and externally?

- **Recovery:** How will you fix the underlying issues, test changes, and restore safe operation?

These questions should be captured in AI-specific runbooks, not improvised in the middle of a crisis.

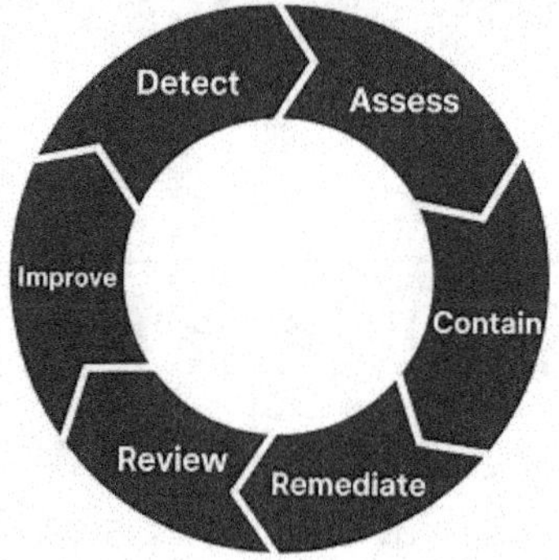

3.7 Safety roles and responsibilities

Safety fails when everyone assumes someone else is responsible. As part of your AI modernization governance, you should explicitly define safety-related responsibilities:

- **Product owners** define risk levels, safety requirements, and escalation paths.
- **Data and ML teams** implement safety controls (filters, thresholds, monitoring) and document limitations.

- **Security and risk teams** evaluate misuse scenarios and integrate AI safety into broader risk management.
- **IT operations** monitor production systems and act on alerts.
- **Leadership** sets risk appetite and approves high-impact deployments.

As the modernization leads, you are responsible for ensuring this safety map exists and is understood.

AI Safety RACI Map

Roles	Risk Assessment	Safety Measures Design	Testing an Validation	Monitoring
AI Ethics Lead	R	C	R	C
AI Engineers	R	C	C	I
Compliance Officer	A	C	C	R
Management	I	C	C	A

Safety controls across different AI use cases

Not all AI systems require the same level of safety engineering. A low-impact internal summarization tool and a high-impact eligibility determination system should not be treated the same. As a manager, you can categorize systems into **safety tiers** and adjust controls accordingly:

- **Tier 1 – Low impact:** Internal productivity tools with limited external effect. Light monitoring, basic guardrails.
- **Tier 2 – Moderate impact:** Customer-facing tools that influence but do not decide outcomes. Stronger filters, human-in-the-loop.
- **Tier 3 – High impact:** Systems that directly affect rights, access, or safety. Rigorous testing, approvals, and continuous monitoring.

This tiered approach focuses your resources where they matter most.

Safety Tier Pyramid

Real-world pattern: misinformation and hallucinations

Generative AI systems can confidently produce incorrect or fabricated information. In some domains, this is inconvenient; in others, it is dangerous. Imagine an AI assistant that occasionally fabricates policy details, medical suggestions, or legal interpretations. Even if these hallucinations are rare, the impact can be severe.

For modernization leaders, safety requires:

- Constraining systems to approved knowledge sources where possible.
- Explicitly labeling AI-generated content and clarifying its role (advisory vs authoritative).
- Implementing confidence thresholds and fallback behaviors when the system is uncertain.
- Training users not to over-trust AI output and building interfaces that encourage verification.

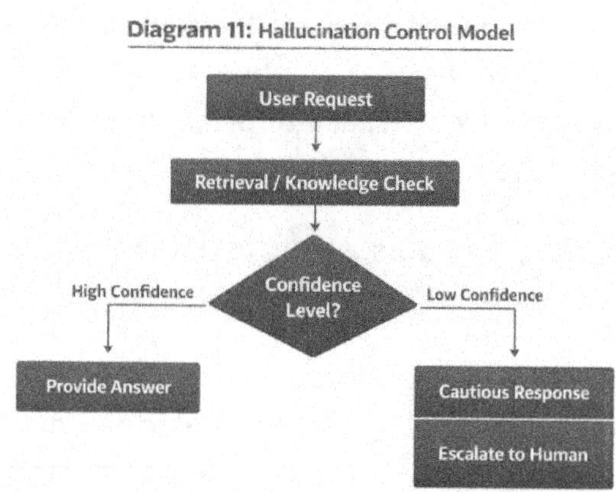

Real-world pattern: automated decision failures

AI is increasingly used to automate or assist high-impact decisions: fraud detection, claim approvals, credit limits, and case prioritization. Decision failures can manifest as false positives (blocking legitimate users) or false negatives (missing harmful activity). Both can be safety issues.

As an IT manager, you must design for:

- Clear boundaries between automated and human decisions.
- Review processes for borderline or ambiguous cases.

- Feedback loops where human reviewers can correct system errors.
- Impact monitoring to identify whether certain groups are disproportionately affected.

Safety here is about distributing risk responsibly, not simply delegating decisions to machines.

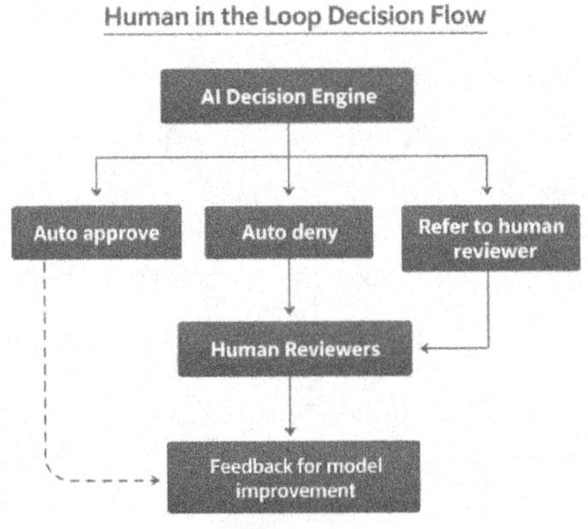

3.8 The Manager's Safety Playbook

To operationalize AI safety in your modernization program, you can adopt a simple, repeatable playbook:

1. **Classify systems by impact.** Assign each AI system to a safety tier based on decision impact and exposure.

2. **Run a harm scenario workshop.** For higher systems, bring cross-functional stakeholders together to identify misuse and failure scenarios.
3. **Define guardrails and thresholds.** Decide what the system is allowed to do, and where it must defer to humans.
4. **Embed monitoring and alerts.** Implement logging, anomaly detection, and clear alert routes for safety-relevant issues.
5. **Create AI incident runbooks.** Document who does what when something goes wrong. Test those plans with tabletop exercises.

This is not a one-time exercise. You revisit these steps as systems evolve, usage scales, and new risks emerge.

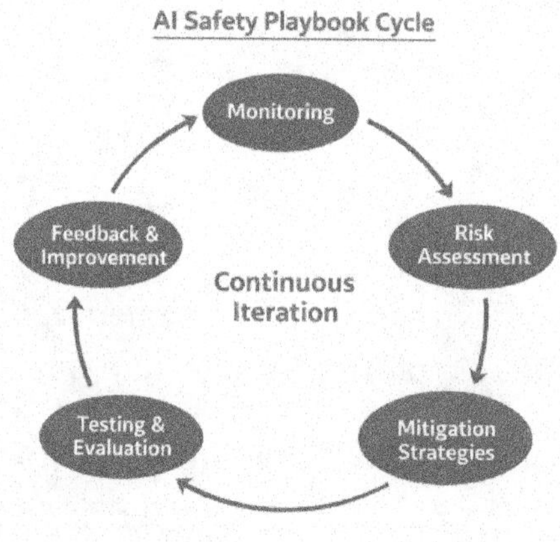

The AI Reality Check: Safety beyond checklists

It is tempting to treat safety as a checklist to satisfy policy or compliance. In reality, safety is a mindset and a practice. Checklists help, but they cannot cover every scenario in a rapidly evolving environment. As an IT manager, you must:

- Encourage teams to speak up when something "feels unsafe," even if it passes formal checks.
- Accept that some high-risk use cases should be delayed or declined entirely.
- Collaborate closely with security, legal, compliance, and business stakeholders rather than keeping AI decisions in an IT silo.

Your job is not to eliminate all risk—that is impossible. Your job is to ensure risks are visible, discussed, and actively managed.

Checklist vs Culture

Both Needed for True AI Safety

The Monday Morning Test

On Monday morning, a safety-aware IT manager should be able to answer:

1. Which AI systems under my responsibility are high-impact and require the strictest safety controls?
2. For those systems, what documented guardrails, thresholds, and escalation paths exist today?
3. How would we detect if one of these systems started causing harm or being misused?
4. Who is on point for AI safety decisions in my organization, and how do I engage them quickly?
5. When was the last time we tested our AI incident response process—and what did we learn?

If some answers are unclear right now, that is your roadmap. Safety is not merely a technical feature; it is a management discipline that will define whether your AI modernization builds durable trust—or fragile, short-lived wins.

AI Safety Readiness Dashboard

Number of High Impact Systems

% with Defined Guardrails

Higesst

Open Safety Issues

Chatbots Recommendation systems

Time Since Last Incident Drill

92 days

4 Fairness at Scale

Opening vignette

You are rolling out an AI model to prioritize service tickets across a nationwide customer base. Early metrics look excellent: faster resolution times, lower backlog, better utilization of senior engineers. Then a regional manager calls with a concern: tickets from certain neighborhoods and smaller customers seem to be consistently pushed to the back of the queue. A few key clients complain they are "being treated as second-class customers by the algorithm." Your COO forwards the email and asks a blunt question: "Is our AI fair—and how would we know?" In that moment, you realize fairness is not a feature you toggle on. It is a continuous practice you must design into your workflows, your data, and your governance.

4.1 Why fairness is a business and mission issue

Fairness is often framed as a purely ethical concern, but for IT managers leading AI modernization, it is also a business and mission issue. Unfair AI systems erode trust, invite regulatory and legal scrutiny, and damage relationships with customers, citizens, and employees. They can quietly sideline valuable segments, misallocate resources, and amplify existing inequities embedded in historical data.

Fair AI, on the other hand, enables sustainable scale. When users believe your systems treat them equitably, they are more willing to adopt new AI-enabled channels, share data, and rely on automated decisions. For modernization leaders, fairness is not just "doing the right thing"; it is protecting your organization's legitimacy and license to operate as you expand AI.

Fairness as Risk and Value

4.2 Fairness is not a one-time configuration

A common misconception is that fairness can be "configured" once—by training on a better dataset or applying a single bias mitigation method—and then forgotten. In reality, fairness is dynamic. Models drift, populations change, and usage patterns evolve. A model that appears fair at deployment can become unfair over time as behaviors, incentives, or data sources shift.

For IT managers, this means that fair work must be systematic and ongoing. It needs to show up at:

- Intake: when you define the problem and stakeholders.
- Design: when you choose data, models, and workflows.
- Testing: when you evaluate performance and fairness metrics.
- Operations: when you monitor, audit, and adjust over time.

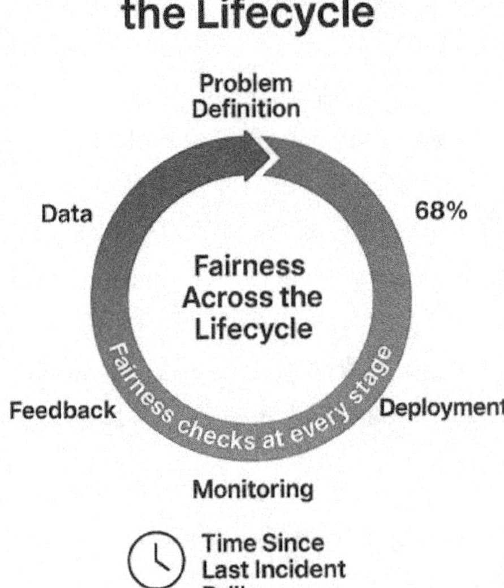

4.3 Understanding common sources of bias

To manage fairness, you need a clear picture of where bias can enter. In practice, bias typically arises from four main sources:

1. **Historical data bias:** Data reflects past decisions, which may have favored or disadvantaged certain

groups (for example, historic underwriting decisions or hiring patterns).

2. **Sampling bias:** Certain groups or scenarios are underrepresented in the training data, so the model performs poorly for them.

3. **Measurement bias:** The labels used (default, fraud, "high potential") are noisy or reflect subjective judgments.

4. **Workflow and interface bias:** Even if the model is balanced, the way its outputs are used in systems can create unfair outcomes (for example, scoring thresholds used differently across regions).

As an IT manager, you will not fix all of these yourself, but you must ensure they are considered by the teams building and deploying AI.

Four Sources of Bias

Historical Data	Sampling
• Past inequities baked into records	• Certain groups underrepresented
Measurement	**Workflow**
• Proxies that miss real need	• Biased decisions in practice

Individual vs group fairness: what you need to know

Fairness literature distinguishes between many formal definitions, but managers need a practical grasp of two main perspectives:

- **Individual fairness:** Similar individuals should receive similar outcomes.
- **Group fairness:** Different demographic groups should have comparable treatment or outcomes (for example, similar error rates or approval rates where appropriate).

You rarely achieve perfect fairness on all definitions simultaneously. Part of your governance role is to work with stakeholders (legal, compliance, policy, HR, business owners) to select fairness goals that fit the context. For example, in hiring, you might emphasize similar selection rates across demographic groups; in fraud detection, you might focus on similar false positive rates across groups.

4.4 The Manager's fairness checklist (project level)

Fair work starts with asking structured questions early in the project. Below is a project-level fairness checklist you can embed into intake and design reviews:

Fairness Project Checklist

- Stakeholders: Who is impacted by this AI system? Which groups might be particularly vulnerable or historically disadvantaged?
- Use suitability: Is it appropriate to use AI here, given the potential for harm if things go wrong?

- Protected characteristics: Which attributes (race, gender, age, disability, income level, location, etc.) are relevant to fairness considerations—even if not directly used as inputs?
- Benefit and burden: Who benefits if the system works well? Who bears the cost when it fails?
- Baseline disparities: Are there existing inequities in the current (non-AI) process that the system might reinforce or mitigate?
- Decision type: Is the AI making decisions, recommendations, or prioritizations that meaningfully affect people's opportunities or treatment?
- Appeal and recourse: How can people affected by decisions challenge them or ask for review?

4.5 Designing fairness checks into workflows

Fairness cannot live only in the model. It must be integrated into the model's workflow. As an IT manager, you can influence fairness by shaping:

- Where AI is allowed to act autonomously vs requiring human review. High-impact or ambiguous cases might always go to a human.
- **How thresholds and business rules are set.** Different thresholds can create different fairness

profiles—sometimes a slightly lower overall accuracy, but a more equitable impact is the right trade-off.

- **What information is exposed to human decision-makers?** Interfaces that show only scores without context can encourage blind trust; interfaces that highlight uncertainty and the fairness implications can support better judgment.
- **How feedback is captured.** Your systems should allow frontline staff and users to flag potential unfairness, which feeds back into model and workflow improvements.

"Workflow-Aware Fairness"

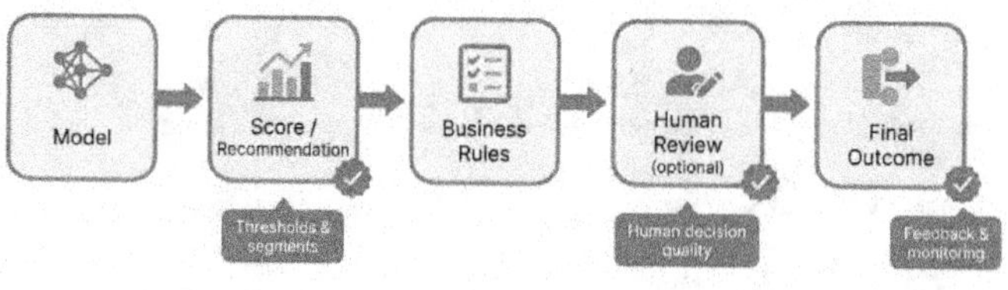

Fairness metrics without math overload

You do not need to derive fair formulas, but you do need to understand what your teams are measuring. Commonly used fairness metrics include:

- **Selection rate comparisons:** Are different groups being selected/approved at similar rates, where appropriate?
- **Error rate parity:** Are false positive and false negative rates similar across groups?
- **Calibration:** For a given risk score, do outcomes (default, fraud, success) occur at similar rates across groups?

Your role is to ask:

- Which fairness metrics are we using and why?
- What tradeoffs are we making between fairness and other goals?
- How will we monitor these metrics over time in production?

Sample Fairness Metrics Dashboard

Group	Approval Rate	False Positive Rate
Group A	82%	● Fair
Group B	3%	● Mixed
Group C	7%	● Unfair
Feedback	65%	92 days

The bias audit: making fairness measurable

A **bias audit** is a structured process that evaluates whether an AI system exhibits unfair patterns across defined groups. As AI scales, bias audits become a core governance mechanism—similar to security audits or financial audits. For IT managers, bias audits should be:

- **Repeatable:** Using consistent methods and metrics.
- **Risk-based:** Applied more intensively to high-impact systems.
- **Documented:** With clear reports that leadership, regulators, and partners can understand.
- **Action-oriented:** Leading to remediation plans, not just reports.

4.6 Bias Audit Checklist

- Scope: Which system, which decisions, and which time period are we auditing?
- Groups: Which groups are we comparing (for example, gender, region, income band) and why?
- Metrics: Which fairness metrics are we using (selection rates, error rates, calibration, etc.)?
- Thresholds: What level of disparity triggers concern or action?
- Root-cause analysis: For identified disparities, what hypotheses do we have about causes (data, model, workflow)?

- Actions: What concrete changes will we test (data balancing, threshold changes, model retraining, interface changes)?
- Follow-up: When will we re-audit to see if disparities improved?

Bias Audit Pipeline

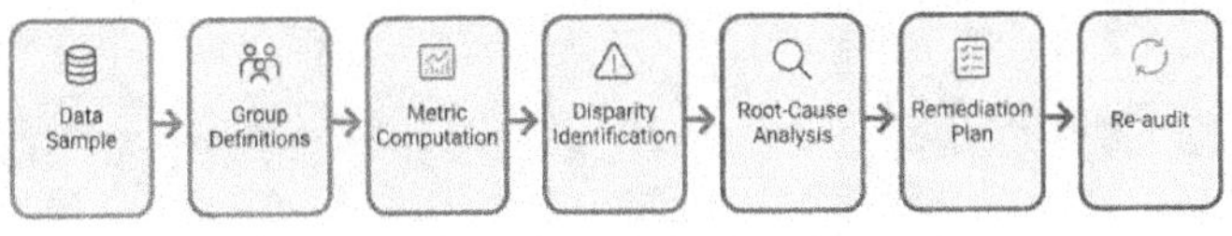

Equity dashboards: making fairness visible over time

One-off audits are not enough. To manage fairness at scale, you need ongoing visibility—where **equity dashboards** come in. An equity dashboard tracks fairness-relevant metrics (such as approval rates, rejection reasons, escalation rates, and error patterns) across different groups and over time.

For IT managers, an effective equity dashboard:

- Focuses on a small number of meaningful metrics, not dozens of charts.
- Is integrated into existing reporting (risk, operations, customer experience), not siloed.

- Supports drill-down from organization-wide trends to specific segments or products.
- Triggers conversations: it raises questions when disparities appear, rather than being a passive report.

Equity Dashboard Layout

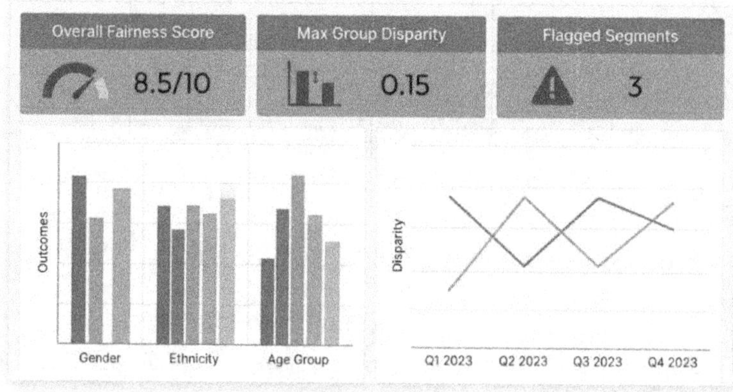

The manager's fairness checklist (operational level)

Once systems are live, your fairness responsibilities shift from design to operations. Use an operational fairness checklist to guide your oversight:

Operational Fairness Checklist

- Monitoring: Do we have production fairness metrics tracked regularly (for example, monthly or quarterly)?
- Alerts: Are there thresholds or rules that flag significant increases in disparity for review?
- Governance cadence: Is fairness a standing agenda item in relevant governance forums (AI review board, risk committee, IT governance)?
- Drift detection: Are we monitoring for changes in data distributions or performance that might impact fairness?
- Feedback: How do we collect and triage fairness-related complaints from users, staff, or external stakeholders?
- Documentation updates: When models, data, or workflows change, do we update fairness assessments and documentation?
- Accountability: Who is responsible for acting when fairness issues are identified?

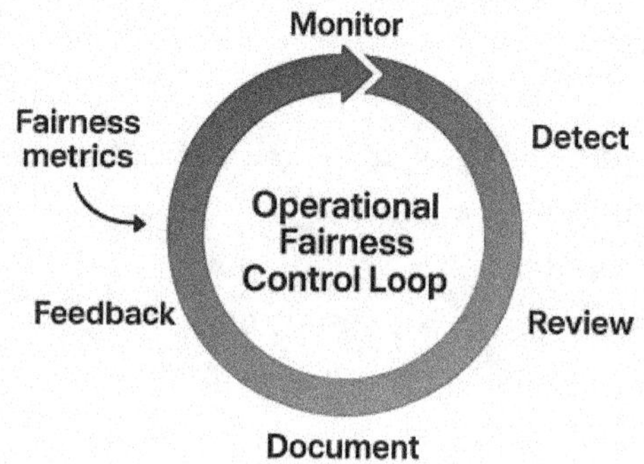

Real-world pattern: fairness in hiring and promotions

AI is increasingly used to screen candidates, rank resumes, and flag internal employees for promotion or development programs. If these systems learn from past decisions, they risk enshrining historical inequities. For example, if previous leadership roles have been disproportionately held by one demographic group, the model may favor profiles like those of those leaders.

As a modernization leader, fairness in this context requires:

- Diverse and representative training data, or at least a clear awareness of limitations.
- Explicit fairness goals (for example, reducing disparities in interview invitation rates).
- Human-in-the-loop decisions for high-stakes outcomes (offers, promotions).

- Audits to track whether AI-supported processes are narrowing or widening gaps over time.

Hiring Fairness Flow

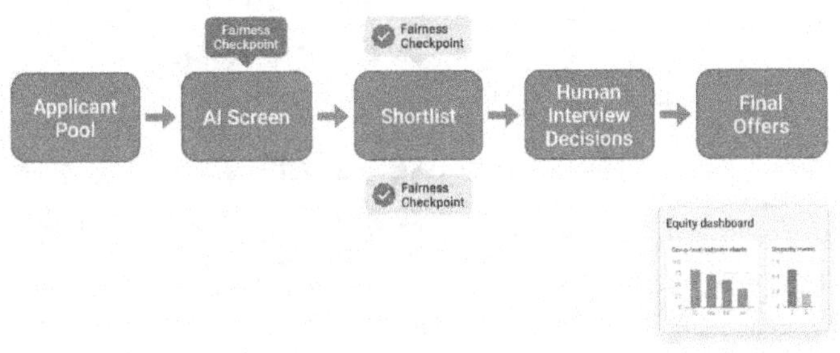

Real-world pattern: fairness in credit, benefits, and risk scoring

In credit scoring, benefits eligibility, insurance pricing, or fraud risk assessment, fairness concerns are acute. Models here directly influence access to essential services and financial opportunities. Bias can show up as certain communities systematically facing higher rejection rates, stricter reviews, or higher prices.

IT managers must ensure that:

- Sensitive attributes (or proxies) are handled carefully in both training and deployment.

- Fairness metrics are tuned to the regulatory and policy context (for example, focusing on unjustified disparities in approval rates and error rates).
- Business stakeholders understand trade-offs between risk management and fairness.
- Appeals mechanisms and human review are available, especially for adverse decisions.

Fairness Risk Heatmap

	Credit	Benefits	Risk Scoring	Marketing Offers
Access Disparity	⚠			
Pricing Disparity			⚠	
False Positives				
False Negatives	⚠			

Legend: ▓ Higher fairness risk

4.7 Governance mechanisms that make fairness stick

To sustain fairness at scale, you must embed it into your governance structure—not rely on ad-hoc heroics. Core governance mechanisms include:

- **Fairness policies and standards:** Clear guidelines on what is expected for fairness assessments, documentation, and remediation.
- **AI review boards:** Cross-functional bodies (IT, data, legal, compliance, HR, operations) that review fairness for high-impact systems.
- **Risk classification:** Categorizing AI systems by fairness risk level and aligning control requirements accordingly.
- **Audit schedules:** Planned, periodic fairness reviews for high-risk systems, with documented outcomes.

Your role is to ensure these mechanisms exist, are used, and have authority.

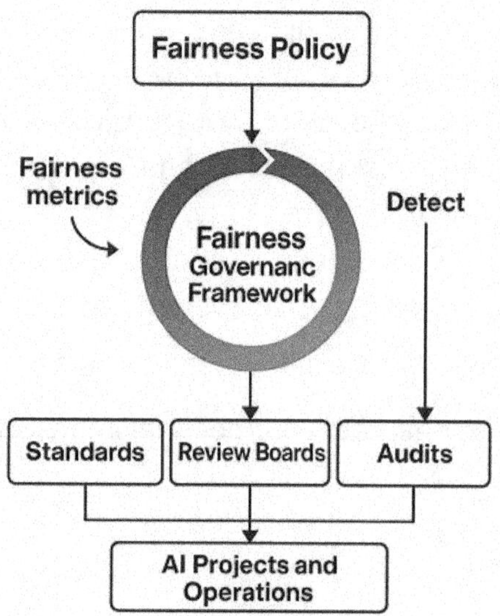

4.8 The Manager's Fairness Playbook (90-day plan)

To move from awareness to action, you can adopt a focused 90-day fairness playbook for your AI modernization program:

0–30 days

- Identify your top 3–5 AI systems with the highest potential fairness impact.
- For each, complete the Fairness Project Checklist retrospectively.
- Align with legal/compliance on fairness expectations and risk appetite.

31–60 days

- Pilot a bias audit on at least one high-impact system.
- Define a minimal equity dashboard for that system (2–3 key fairness metrics).
- Incorporate fairness questions into your standard architecture or design review templates.

61–90 days

- Present findings and remediation plans from the bias audit to leadership and governance forums.
- Formalize a simple fairness policy and embed it in your AI governance documents.
- Plan a cadence (for example, quarterly) for fairness audits and dashboard reviews.

90-Day Fairness Playbook Roadmap

Days 0–30	Days 31–60	Days 61–90
Define fairness goals	Run bias audits	Roll out checkpoints
Inventory high-impact use cases	Design mitigation tactics	Integrate dashboards
Set governance roles	Pilot new workflows	Capture feedback & iterate

The AI Reality Check: avoiding "fairness theater."

It is easy to fall into "fairness theater"—talking about fairness, creating slide decks, or piloting one-off tools without changing decisions or outcomes. As an IT manager, you can spot fairness theater when:

- Fairness discussions never influence project prioritization or go/no-go decisions.
- No metrics or dashboards exist to track fairness in production.
- Bias audits happen once for a pilot and never again at scale.
- People cannot name who is accountable when fairness issues are found.

Real fairness work may slow down some launches, change some designs, and require uncomfortable conversations.

That is the point: fairness that never challenges decisions is probably not real.

Fairness Theater vs Fairness Practice

4.9 The Monday Morning Test

By the time you finish this chapter, you should be able to sit down on Monday morning and answer:

1. Which AI systems in my portfolio have the highest fairness impact, and why?
2. For those systems, which groups are we explicitly monitoring for fairness, and which metrics are we using?
3. Where, in our workflows, are fairness checks concretely implemented (for example, thresholds, human review, dashboards)?
4. When was the last time we assessed bias for a high-impact system, and what actions followed?

5. Who in my organization is accountable for fair decisions, and how do I engage them when issues arise?

If several answers are "I am not sure," that is not a failure; it is a starting point. Fairness at scale is a journey—one that requires intentional design, persistent monitoring, and governance that treats fairness as a first-class objective, not a side constraint.

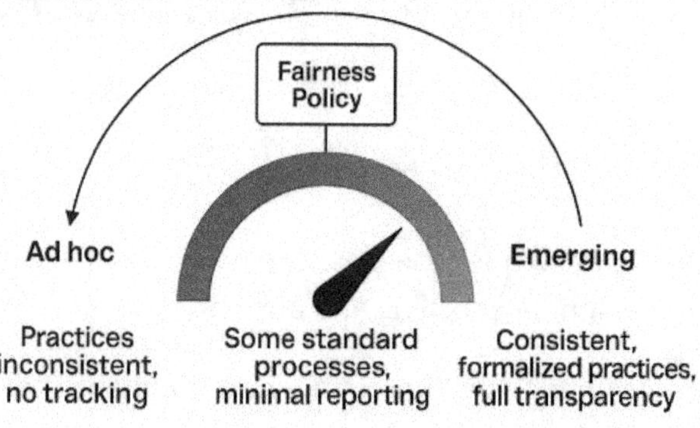

5 Transparency: Seeing Inside the Black Box

Opening vignette

You have just presented the results of a new AI-powered claims triage system to your executive team. Processing time is down 40 percent, staff workload is lower, and backlog is shrinking. The COO is impressed. Then the General Counsel leans in: "Walk me through how this model decides which claims to fast-track and which to delay. If a customer challenges a decision, what can we actually show them?" The room quiets. Your data science lead starts explaining gradient boosting trees and feature importance plots. The executives are not reassured. In that moment, you realize it is not enough that the AI works. People must understand it—at the right level, in the right language—if they are going to trust it. Transparency has become a leadership requirement, not a technical luxury.

5.1 Why transparency is the foundation of accountability

Transparency and accountability go together. You cannot hold anyone accountable for AI-driven outcomes if you cannot see how decisions were made, what data was used, and who approved what. For IT managers leading AI modernization, transparency is what turns AI from a mysterious "black box" into an auditable, governable system.

Without transparency, you face:

- Difficulty explaining decisions to customers, regulators, and internal stakeholders.
- Increased risk of hidden bias, safety issues, or policy violations going unnoticed.
- Tension between IT, business, legal, and compliance because nobody shares a common picture of how AI is operating.

With transparency, you gain:

- Clear lines of responsibility for decisions.
- Faster resolution of incidents and complaints.
- Stronger confidence from leadership to expand AI into critical workflows.

Transparency → Accountability → Trust

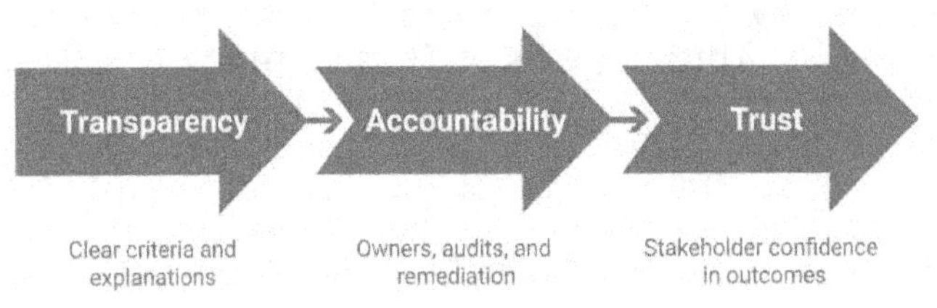

What transparency means in practice (without reading code)

Many managers hear "explainability" and assume it means understanding complex math or reading model code. For modernization leaders, transparency is more practical and layered. You need:

- A **business-level explanation** of what the system does, why it exists, and where its limits are.
- A **process-level view** of how inputs are transformed into decisions and where humans step in.
- An **audit-level record** of specific decisions: what came in, what the system did, and who approved or overrode it.

You are not trying to become a data scientist. You are building a consistent way for your organization to answer the question: "How did this AI affect this decision at this time?"

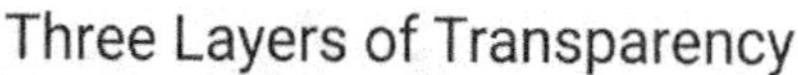

Three Layers of Transparency

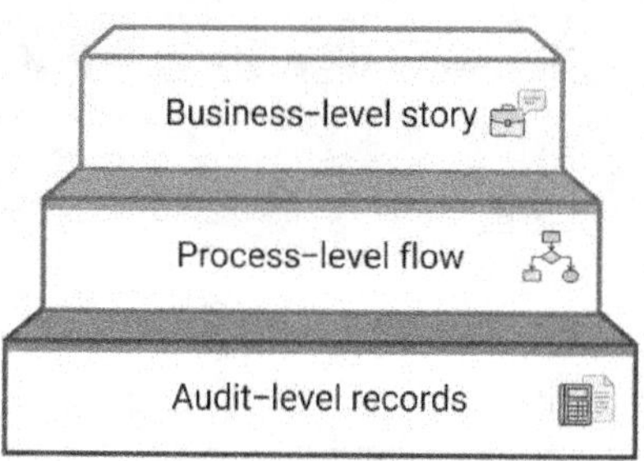

5.2 The building blocks of AI transparency

For AI systems in your modernization portfolio, practical transparency typically rests on four building blocks:

1. **Model documentation** – What is this model? What is it for? How was it built and tested?
2. **Decision logs and audit trails** – What did this system actually do over time?
3. **Explainability tools and views** – How can we interpret specific predictions or decisions?
4. **Communication artifacts** – How do we explain the system to different audiences: users, executives, regulators, and staff?

Your job is to ensure that each important AI system has these building blocks in place before it becomes business-critical.

5.3 Model cards: the "spec sheet" for AI models

A **model card** is like a spec sheet or product label for an AI model. It summarizes key information in a structured, readable way so that managers, risk teams, and other stakeholders can quickly understand what they are dealing with.

A practical model card for your organization might include:

- Model name and version
- Purpose and intended use
- Out-of-scope or prohibited uses
- Data sources and time ranges
- Performance metrics (overall and for key segments)
- Known limitations and risks
- Fairness and safety notes
- Monitoring and retraining plan
- Owners and contacts

You do not need to write the model card yourself, but you should insist that every significant model has one—and that it is kept up to date.

Model Card Layout

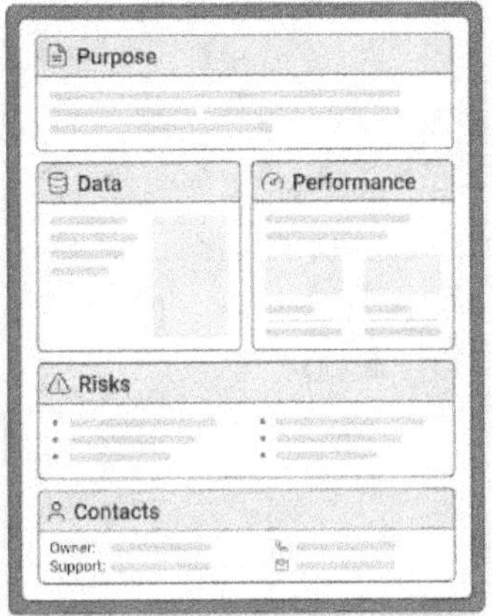

Template: Model card (manager-ready fields)

Below is a fill-in-the-blank model card template you can standardize across your teams:

Model Card – Manager-Ready Template

- Model name:
- Version/release date:
- Business owner:
- Technical owner:

1. Purpose and scope

- Primary purpose of the model:
- Business processes or products it supports:

- Intended users (internal roles):
- Intended decision type (advisory / assistive/automated):

2. Intended use and limits

- Intended use cases (bullets):
- Known out-of-scope or prohibited uses:
- Assumptions about context (data quality, workflows, user behavior):

3. Data overview

- Main data sources:
- Time period covered:
- Key features/attributes used:
- Known limitations or gaps in data:

4. Performance and fairness

- Key performance metrics (e.g., accuracy, AUC, etc.):
- Performance by major segments or groups:
- Known fairness or bias risks:

5. Risks, controls, and safeguards

- Top 3–5 identified risks (e.g., safety, fairness, misuse):
- Key safeguards (guardrails, thresholds, human review steps):

6. Monitoring and lifecycle

- Monitoring signals (what we track in production):
- Retraining or review schedule:
- Change approval process:

7. Contacts and accountability

- Primary contact (business):
- Primary contact (technical):
- Governance forums where this model is reviewed:

Model Card as a Governance Artifact

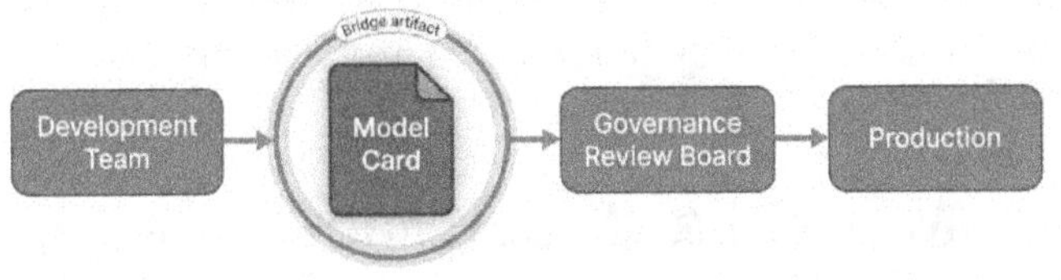

5.4 AI fact sheets: broadening beyond the model

Where a model card focuses on a specific model, an **AI fact sheet** covers the end-to-end AI service or capability as it is deployed in your environment. Think of it as a product sheet for the whole AI-enabled system, not just the underlying model.

An AI fact sheet might include:

- The overall service description and key user journeys.
- All models involved (including third-party or vendor models).
- Data flows: what data goes in and out, and where it is stored.
- Interfaces and integrations with other systems.
- User-facing transparency elements (disclosures, help text, FAQs).
- Compliance and policy alignment notes.

Fact sheets are especially useful for vendor-supplied solutions where you do not control the entire stack but still need transparency for governance and oversight.

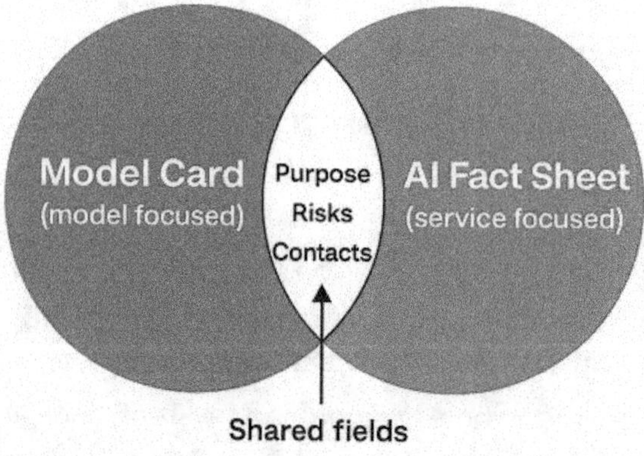

Template: AI fact sheet (system level)

AI Fact Sheet – System-Level Template

1. System overview

- System name:
- Business owner:
- Technical owner:
- Summary of what this AI system does (2–3 sentences):

2. Components and architecture

- List of models used (internal and external):
- Key data sources and sinks:
- Major integrations (internal systems/APIs):

3. User journeys and decisions

- Primary user roles (internal/external):
- Key workflows and decisions influenced by AI:
- Where human review is required or recommended:

4. Transparency and user information

- What users are told about AI involvement (disclosures):
- Links to help pages, FAQs, or guides that explain the system:
- Options for users to request human oversight or review:

5. Risk and compliance alignment

- Regulatory or policy requirements this system touches (high level):
- Known high-risk areas (safety, fairness, privacy):
- Key controls in place:

6. Logging, monitoring, and audit

- Logs captured (inputs, outputs, overrides, errors):
- Retention period and access controls:
- How logs can be exported for audit or regulatory requests:

7. Ownership and governance

- Governance forums where this system is reviewed:
- Change management and approval process for major updates:

Fact Sheet as a Single Source of Truth

Decision logs and audit trails: remembering what the system actually did

Transparency is not just about documentation at design time. It is also about **what actually happened in production**. Decision logs and audit trails give you that record. At a minimum, for consequential AI systems, your logs should be captured:

- Unique request ID and timestamp.
- Key input fields (or a representative summary where privacy requires).
- Model version used.
- Model output (score, classification, generated content).
- Business rules applied after the model's output.
- Final decision or action taken.
- Whether a human overrode or modified the decision.
- Any error or exception information.

When an issue arises—like a complaint about a particular decision or a suspected pattern of bias—these logs become your primary transparency tool.

Decision Log Schema Overview

Decision ID	Date	Owner	Context / Use Case	Options Considered	Final Decision	Rationale	Follow-up Actions
[illegible]	[illegible]	[illegible]	Action text dummy text	[illegible]	Options decision	[illegible]	[illegible]
[illegible]	[illegible]	[illegible]	Plan text to dummy text	[illegible]	Final decision	[illegible]	[illegible]
[illegible]	[illegible]	[illegible]	[illegible]	[illegible]	None	[illegible]	[illegible]

5.5 Manager checklist: transparency and logging

Use this checklist to assess whether an AI system under your responsibility meets a minimum transparency standard:

Transparency & Logging Checklist

- Documentation
    - The system has an up-to-date model card.
    - The system has an AI fact sheet covering end-to-end behavior.
- Logging
    - Key inputs, outputs, and decisions are logged with clear identifiers.

- o Model version and configuration are recorded per decision.
  - o Human overrides or manual decisions are logged distinctly.
- Access and retention
  - o Logs are stored securely with controlled access.
  - o Retention periods align with legal and policy requirements.
  - o Logs can be exported in a structured format for audit.
- Usability
  - o There is a documented process for retrieving and interpreting logs when complaints or incidents arise.

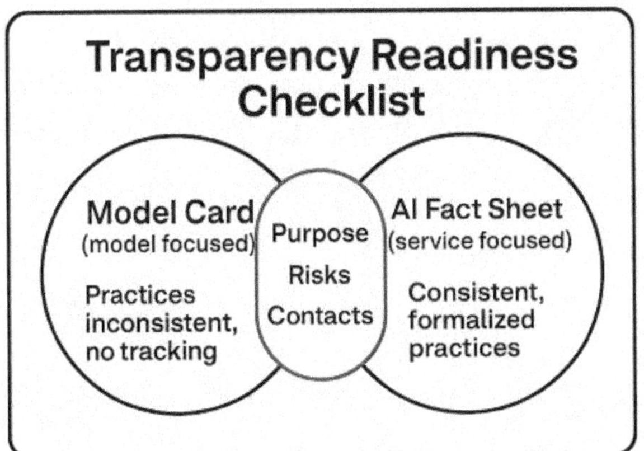

Explainability views: right explanations for the right audience

Explainability tools (like feature importance, example-based explanations, or rule approximations) are powerful, but they can overwhelm non-technical stakeholders if not presented carefully. As an IT manager, you should focus on **audience-appropriate explainability**:

- For executives: high-level drivers of decisions, key limitations, and risk points.
- For frontline staff: clear guidance on when to trust the AI and when to escalate.
- For technical reviewers: more detailed metrics and feature attributions.
- For external stakeholders: plain-language explanations and examples.

You do not need to choose one explanation method. You need layered explanations that match the decision's impact and the audience's needs.

Explainability by Audience

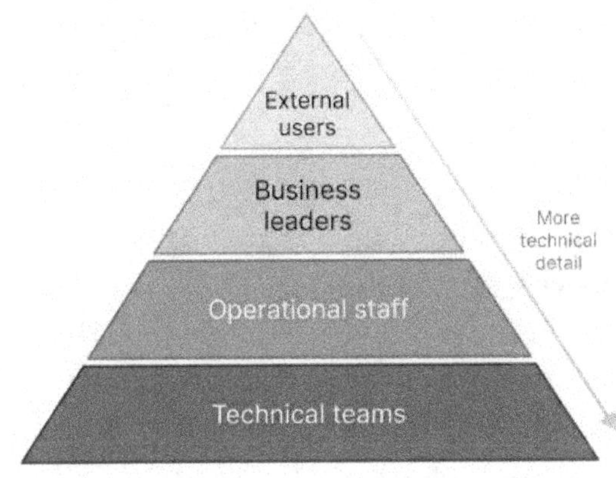

5.6 Communication strategies: how you talk about AI

Transparent communication is as important as technical explainability. If people do not understand what the AI does and does not do, they will either distrust it or over-trust it. Both are dangerous.

As modernization leads, you should ensure that each AI system has:

- A clear, plain-language description of its purpose and limitations.

- Standard language for disclosures ("An AI system supported this decision. A human can review it at your request.").
- FAQs or help content that address common questions and concerns.
- Internal training materials so staff know how to interpret and talk about AI decisions.

Your job is to prevent surprises. Stakeholders should never feel they "discover" AI after the fact.

AI Communication Map

	UI text	Training	Reports	Briefings
Customers	●		○ ○	
Employees	●	●	●	○
Executives			●	●
Regulators			●	●

Template: AI decision notice (external communication)

For systems that significantly affect users, you can standardize a short decision notice. Adapt this to your domain and regulatory context:

5.7 AI-Supported Decision Notice – Template

- **What happened:**
  "This decision about [topic: application, claim, request] was supported by an AI system used by [organization]."
- **What the AI did:**
  "The system analyzed [types of information used, in plain language] and provided a recommendation to our staff."
- **Human involvement:**
  "A human [reviewed/confirmed/made the final decision based on] this recommendation."
- **Limitations:**
  "The AI system does not consider [list key things it does not see, if relevant]. It may not always get it right, and we continuously review and improve its performance."
- **Your options:**
  "If you have questions or would like a human to review this decision, please contact [channel] within [timeframe]."

Decision Notice Layout

Decision Notice Layout	
Subject:	
Decision:	☐ Purpose ☐ Risks ☐ Decisions communicated ☐ Responsibilities defined

Templates: internal transparency brief for executives

For each major AI initiative, create a concise executive briefing that answers the questions leaders actually ask.

5.8 Executive AI Transparency Brief – Template

- System name and purpose (2–3 sentences).
- Where it is used (business processes, regions, user groups).
- What decisions it influences or makes.
- Key performance metrics (in business terms).
- Top 3 risks (fairness, safety, misuse, compliance) and how you are controlling them.

- Summary of transparency measures (model card, fact sheet, logging, user disclosures).
- Who is accountable (name and role).

This brief becomes your go-to artifact whenever questions arise at the leadership table.

Executive Brief Snapshot

Context & Objective	Key Risks & Mitigations

Key Metrics & Trends

Decisions & Owners	Next 30 Days

The Manager's Transparency Playbook (60-day plan)

To embed transparency in your AI modernization work, use this short playbook.

Days 0–30

- Inventory AI systems in production or late-stage pilots.
- Prioritize the top 3–5 systems by impact and risk.

- For each, ensure that at least a minimal model card exists.

Days 31–60

- Create AI fact sheets for the top systems, focusing on end-to-end behavior.
- Implement or improve decision logs with proper identifiers and model versioning.
- Roll out the transparency and logging checklist into your governance reviews.

Optional: pilot the decision notice and executive brief templates for one high-impact system.

Transparency Playbook Roadmap

Days 0–30	Days 31–60
Map key decisions	Draft templates & fact sheets
Identify stakeholders	Pilot explainability content
Define transparency levels	Review with governance

5.9 The AI Reality Check: avoiding "black-box comfort."

It is easy to slide into "black-box comfort": trusting that as long as metrics look good, you do not need to understand how AI works. That comfort is fragile. It breaks the moment something goes wrong, and someone asks for an explanation you cannot provide.

As an IT manager, you avoid this trap by:

- Refusing to deploy high-impact AI without baseline transparency artifacts.
- Asking teams to show—not just tell—how decisions can be traced and explained.
- Treating transparency gaps as risks, not inconveniences.

If you cannot explain it, you cannot defend it. Moreover, if you cannot defend it, it should not be running unsupervised in your production environment.

Black Box Comfort vs Transparent Control

Black Box Comfort	Transparent Control
Comfort now, crisis later	Effort now, control later

The Monday Morning Test

On Monday morning, a transparency-focused IT manager should be able to answer:

1. For each high-impact AI system I am responsible for, can I access an up-to-date model card and AI fact sheet within minutes?
2. If a regulator, executive, or customer asks, "Why was this decision made?", can I retrieve logs that show inputs, model version, and decision path for a specific case?

3. Do my teams know which transparency templates and checklists to use for new AI projects?
4. Are explanations tailored for different audiences, or are we trying to reuse the same technical explanation for everyone?
5. When was the last time we reviewed transparency gaps as a risk topic in a governance forum—and what changed as a result?

If some answers are "not yet," that is the signal to act. Transparency is not about perfection; it is about having the structures and artifacts that let you see inside the black box when it matters most, and lead your AI modernization with clarity rather than guesswork.

Transparency Maturity Gauge

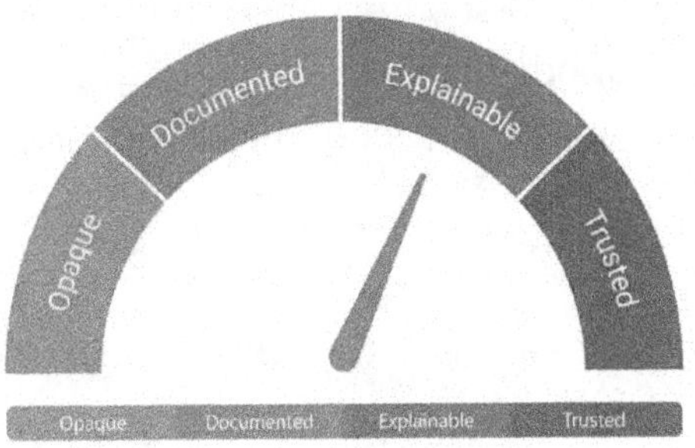

Use this to self-assess current transparency practices

6 The Accountability Chain

Opening vignette

You lead an AI modernization program that includes a new model to prioritize fraud investigations. It has been live for six weeks when a major incident hits: a long-standing customer has their account frozen based on an AI flag that turns out to be wrong. Operations blame the model. The data science team says the business set thresholds. Compliance asks why the model was approved without their sign-off. The COO turns to you: "Who owns this? Who is responsible for making this right—and for making sure it does not happen again?" In that moment, you see the gap: everyone touches the AI system, but nobody clearly owns the consequences. The missing link is an explicit accountability chain.

6.1 Why accountability is the backbone of AI modernization

In traditional IT, accountability often followed infrastructure: the database team owned data availability, the network team owned connectivity, and the application team owned uptime. With AI, accountability shifts to **outcomes**: who is responsible for decisions that affect customers, citizens, or employees when algorithms shape those decisions?

Without a clear accountability chain, organizations experience:

- Finger-pointing when AI-driven decisions go wrong.
- Slow, confused responses to incidents and complaints.
- Governance meetings where issues are logged but not owned.

With a defined accountability chain, you get:

- Clarity on who approves, who operates, and who escalates.
- Faster resolution of issues and clearer remediation paths.
- Confidence from leadership that AI is manageable—not a loose cannon.

Why Accountability Matters

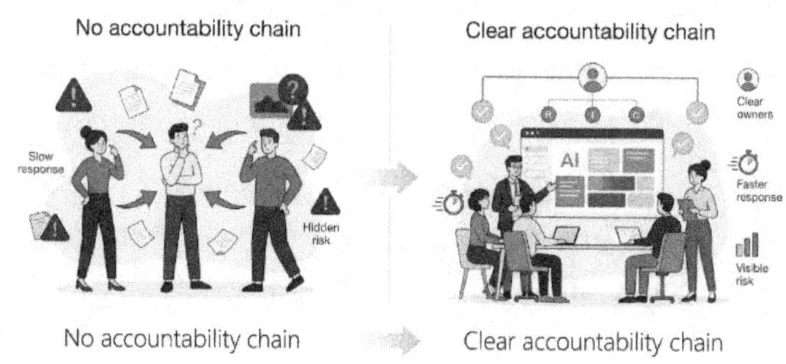

No accountability chain → Clear accountability chain

6.2 Accountability across the AI lifecycle

Accountability is not a single role; it is a chain that runs across the AI lifecycle. For each phase—data, model, deployment, and operations—someone must be clearly accountable (not just "involved"). As the IT modernization leads, your job is to make that chain explicit and visible.

At a high level, accountability spans:

- **Data collection and preparation:** Who owns the datasets, their quality, and their lawful use?
- **Model design and training:** Who owns model behavior, limitations, and risk assessments?
- **Integration and deployment:** Who decides where and how the model is used in workflows?
- **Operations and monitoring:** Who owns outcomes in production and incident handling?

Accountabiliity Across the Lifecy cle

Black Box Comfort   **Transparent Control**

Data	Model	Deployment
Accountable role(s)		Accountable role(s)
Comfort now, crisis later		Effort now, control later

6.3 Mapping the Accountability Chain: roles you must define

To create a real accountability chain, start by naming the key roles (not job titles, but responsibilities). At a minimum, most organizations need:

- **Business owner:** Accountable business outcomes, risk appetite, and using the AI in a specific domain.

- **Data owner:** Accountable for data sources, quality, and lawful use.
- **Model owner:** Accountable for model performance, documentation, and limitations.
- **System owner (IT):** Accountable for integration, deployment, and technical operations.
- **Risk/compliance lead:** Accountable for alignment with policies and regulations.
- **Executive sponsor:** Accountable for high-impact decisions and escalation outcomes.

You may combine some roles in smaller organizations, but you must not leave any unassigned.

AI Accountability Chain Overview

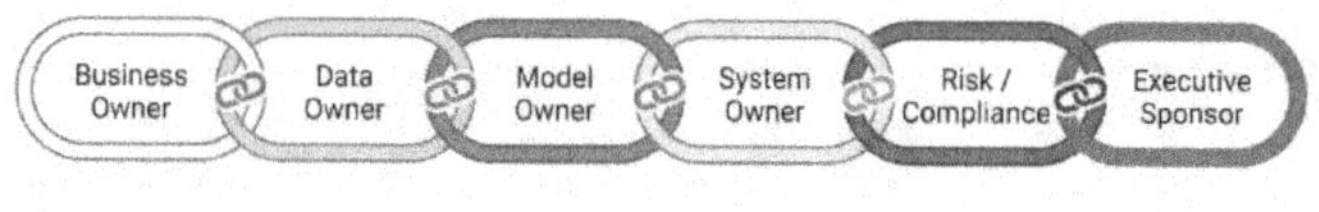

Template: Accountability Chain Mapping (per system)

Use this template for each significant AI system in your portfolio. It can fit on a single page.

AI System Accountability Chain – Template

- System name:

- Business process/area:

1. Business accountability

- Business owner (name, role):
- Accountable business KPIs:

2. Data accountability

- Primary data owner(s):
- Key datasets and systems of record:
- Data quality obligations (e.g., update frequency, validation responsibilities):

3. Model accountability

- Model owner (name, role):
- Model scope (what it is allowed to decide or recommend):
- Known limitations documented:

4. Technical/operational accountability

- System owner (IT):
- Environment(s) where deployed (test, staging, production):
- Uptime, performance, and integration responsibilities:

5. Risk and compliance accountability

- Risk/compliance contact:

- Required approvals (policies, regulatory constraints):

6. Executive accountability

- Executive sponsor:
- Escalation to:

6.4 Decision rights: who decides what, and when

Accountability is unclear when decision rights are fuzzy. For each AI system, you must define **who has the authority to make which decisions**—both during build and in production. Key decision rights include:

- **Go/no-go for deployment:** Who signs off that the system is ready to go live?
- **Risk acceptance:** Who can accept residual risks (e.g., known limitations, partial fairness gaps)?
- **Configuration and thresholds:** Who sets thresholds for scores, alerts, or auto-approvals?
- **Change control:** Who approves major changes (new data, new model version, new use case)?
- **Kill switch:** Who can temporarily suspend or roll back the system when issues arise?

As the IT manager, you do not need to hold all rights, but you must ensure they are clearly defined and documented.

Decision Rights Matrix

	Business owner	Model owner	System owner	Risk	Executive
Deploy	A	C	C	C	A
Risk accept	C	C		A	A
Change	A	A	C	C	
Kill switch	C	C	C	A	A

6.5 Approval checkpoints: building control into the flow

Approval should not be a mystery. A clear **approval-checkpoint framework gives teams a predictable path for** building, reviewing, and deploying AI systems. A simple pattern that works for many organizations includes four checkpoints:

1. **Concept approval:** Business and IT agree that the use case is appropriate for AI and aligned with the strategy.
2. **Pre-deployment review:** Technical, data, risk, and compliance teams review model cards, fact sheets, and test results.
3. **Go-live approval:** A named approver (or committee) signs off that pre-deployment conditions are met.

4. **Post-deployment review:** After an initial period, the system is reviewed against real-world performance, fairness, and incident data.

Approval Checkpoint Flow

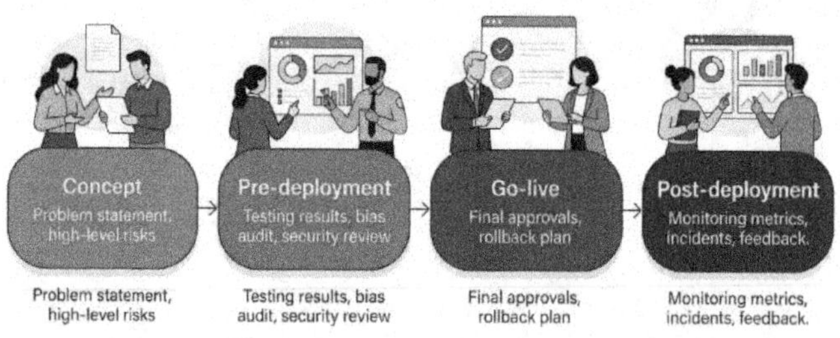

Problem statement, high-level risks

Testing results, bias audit, security review

Final approvals, rollback plan

Monitoring metrics, incidents, feedback.

Template: AI approval checklist (per checkpoint)

Use this condensed checklist at each checkpoint so approvals are consistent and auditable.

AI Approval Checklist – Template

At checkpoint: [Concept / Pre-deployment / Go-live / Post-deployment]

- Scope and use
  - Use case defined and in scope for AI.
  - Stakeholders and impacted groups identified.
- Risk and controls
  - Key risks (safety, fairness, misuse, privacy) assessed.

- o Required controls and guardrails documented.
- Documentation
  - o Model card completed and reviewed.
  - o AI fact sheet completed and reviewed.
- Testing and validation
  - o Performance validated on relevant data segments.
  - o Fairness and bias assessment completed (as required by risk tier).
  - o Security and misuse scenarios considered.
- Operations and monitoring
  - o Logging and monitoring plan in place.
  - o Incident and escalation plan defined.
- Approvals
  - o Approvers for this checkpoint identified.
  - o Decision recorded (Approved / Conditional / Not approved), with rationale.

Approval Checklist Card

Approval Checklist Card	
Purpose ☐ Clear objectives ☐ Defined measures	**Oversight** ☐ Roles identified ☐ Duties assigned
Aiscunts ☐ Assumptions noted ☐ Challenges assessed	**Risks** ☐ Assumptions noted ☐ Challenges assessed

6.6 Escalation paths: what happens when something goes wrong

An accountability chain must include a clear **escalation path** for problems. When a serious issue arises—like a suspected unfair pattern, a harmful recommendation, or a misuse incident—your teams should not improvise who to call.

A practical escalation structure for AI systems includes:

- **Operational escalation:** First line (operations team/system owner) handles immediate containment and logging.

- **Risk and compliance escalation:** When issues have safety, fairness, legal, or reputational implications.
- **Executive escalation:** For incidents that cross defined impact thresholds (volume, severity, or public exposure).

You should document specific escalation triggers—by severity, type of harm, or affected population.

AI Escalation Ladder

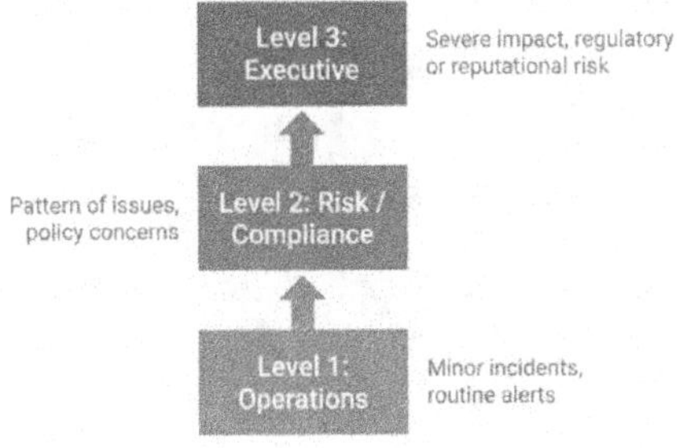

Template: AI incident escalation plan (per system)

AI Incident Escalation Plan – Template

- System name:
- System owner (IT):

1. Incident types and severity levels

- Type A: Operational issues (errors, downtime, integration failures).
- Type B: Safety/fairness issues (harmful decisions, fairness anomalies, misuse).
- Type C: Data/privacy issues (unexpected data exposure or use).
- Severity levels: 1 (low), 2 (moderate), 3 (high), 4 (critical).

2. Escalation triggers

- For each type and severity, define:
    - Who must be notified.
    - How quickly.
    - Through which channel (ticket, email, paging, etc.).

3. Roles in incident response

- Incident commander role:
- Technical lead:
- Business lead:
- Communications lead (if needed):

4. Decision rights during incidents

- Who can decide to halt the system temporarily?
- Who can approve workaround changes?

5. Post-incident review

- Forum where the incident is reviewed:

- Timeline for completing root-cause analysis and remediation plan:

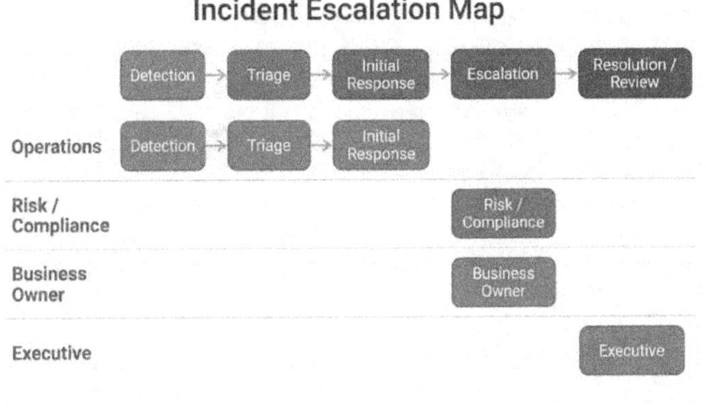

6.7 RACI for AI accountability: clarifying collaboration

A **RACI matrix** (Responsible, Accountable, Consulted, Informed) helps you avoid both overlaps and gaps. Use it to clarify roles for key AI governance activities.

Key activities might include:

- Define use case and requirements.
- Approve data sources and usage.
- Build and validate models.
- Approve deployment and changes.
- Monitor production performance and risk.
- Handle incidents and remediation.

You then assign R/A/C/I across roles (Business owner, Data owner, Model owner, System owner, Risk/Compliance, Executive sponsor).

AI Governance RACI Matrix

	Business Owner	Data Owner	Model Owner	System Owner	Risk / Compliance	Executive Sponsor
Define use case	R			R		R
Approve data sources		A	C	R	I	R
Validate model performance	R	A	C	R	A	
Monitor in production	R		C	C	I	
Handle incidents	R	A		A	I	

R = Responsible; A = Accountable; C = Consulted; I = Informed.

6.8 Manager checklist: Is your accountability chain real?

Use this checklist to stress-test your accountability setup for a given AI system:

Accountability Chain Reality Check

- Ownership
  - We have named individuals for business, data, model, system, risk, and executive roles.
  - Those individuals acknowledge their role and responsibilities.

- Documentation
  - An accountability chain document or template exists and is findable.
  - Decision rights (go/no-go, kill switch, risk acceptance) are documented.
- Approvals
  - The system's deployment history includes clear approval records.
  - Risk acceptance decisions (if any) are formally recorded.
- Incidents
  - There is a documented escalation path for AI incidents.
  - At least one incident drill or tabletop exercise has been run or scheduled.

If you cannot tick these boxes, your accountability chain is more of an aspiration than a reality.

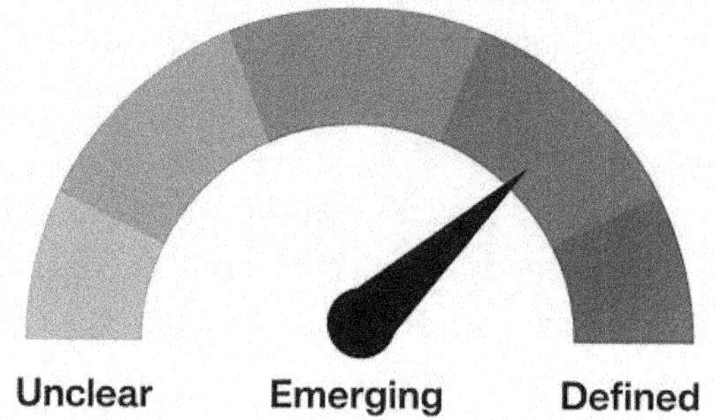

6.9 Real-world pattern: vendor vs internal accountability

Many modernization programs rely on external vendors or cloud services for AI capabilities. When something goes wrong, there is a temptation to blame the vendor. However, from the perspective of regulators, customers, or citizens, **your organization** is accountable for how AI impacts them, even if a supplier is at fault.

For IT managers, this means:

- Contracts must clearly define vendor responsibilities for quality, documentation, and incident support.

- Internally, you still assign business, system, and risk owners for vendor-powered systems.
- Vendor model cards or documentation must be incorporated into your own fact sheets and accountability documents.
- You maintain the right and the mechanism to disable or restrict use if risk exceeds your tolerance.

Shared Accountability with Vendors

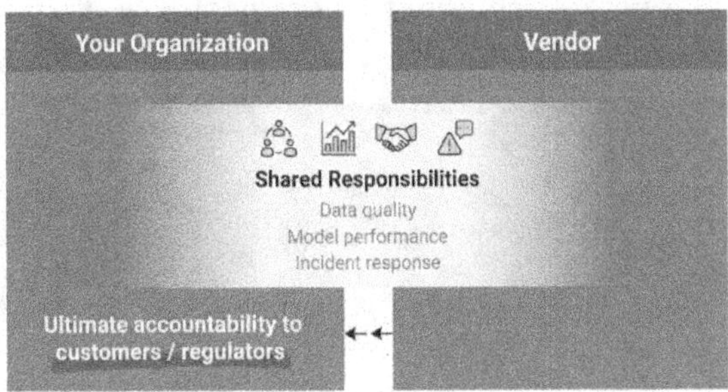

Real-world pattern: continuous learning and drifting accountability

Some AI systems continuously learn or are regularly retrained. Over time, their behavior changes—sometimes in subtle ways. Without a robust accountability chain, these changes can outpace governance. The model owner may assume that changes are minor; the business owner may not realize that behavior has shifted; and nobody revisits risk acceptance.

To prevent this, you should:

- Treat significant retraining as a change that requires reassessment and, for higher-risk systems, fresh approvals.
- Use model versioning tied to approval records and decision logs.
- Set thresholds for drift (performance, fairness, or safety metrics) that trigger review.

Accountability must follow the model over time, not just at its birth.

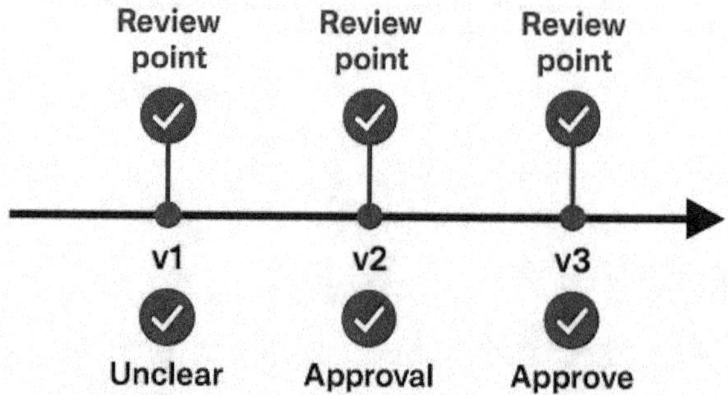

The Manager's Accountability Playbook (90-day plan)

To strengthen accountability in your AI modernization effort, use this 90-day playbook.

0–30 days: Make accountability visible

- Inventory your AI systems and identify the top 3–5 by impact.
- For each, draft an Accountability Chain Mapping using the template.
- Confirm role owners (business, data, model, system, risk, executive).

31–60 days: Formalize decision rights and approvals

- Define decision rights (deploy, kill switch, risk acceptance) for each high-impact system.
- Introduce the AI approval checklist into your governance forums.
- Document existing approvals and any gaps discovered.

61–90 days: Build escalation and test it

- Create an AI incident escalation plan for at least one high-impact system.
- Run a tabletop drill simulating an AI failure or fairness complaint.
- Update accountability documents based on what you learned.

Accountability Playbook Roadmap

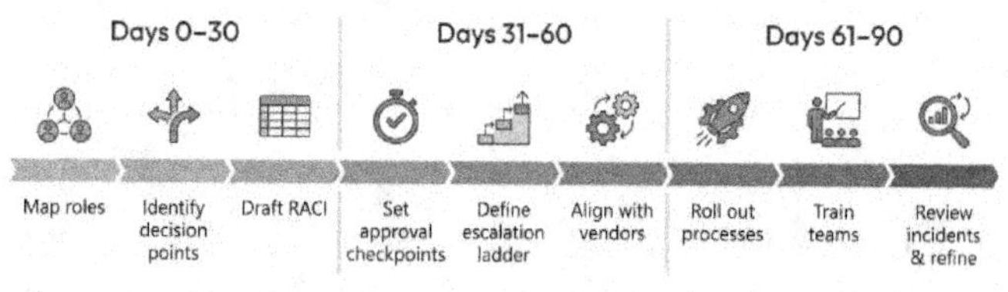

The AI Reality Check: holding systems and humans accountable

There is a risk that organizations try to hold the "AI" accountable, as if the system itself could bear responsibility. That is not how accountability works. AI is a tool. Humans and organizations decide where to use it, how to design it, and how to respond when it fails.

As an IT manager, your test is simple:

- When an AI-influenced decision harms someone, can you name the accountable person or role at each step of the chain?
- Are those people empowered to act—and expected to?

If not, you have governance theater, not governance. The accountability chain is what turns AI from "something we deployed" into "something we are responsible for."

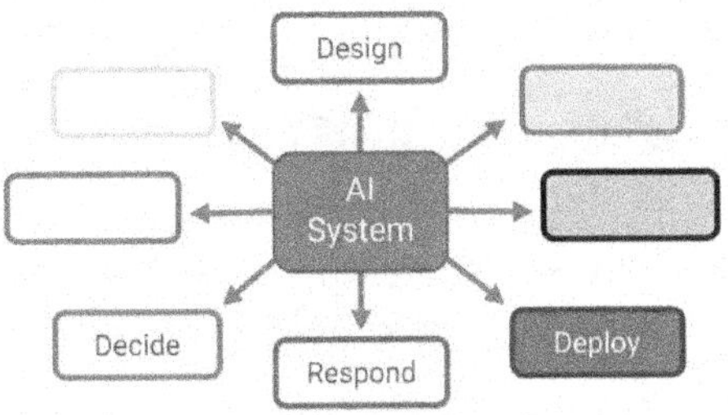

The Monday Morning Test

On Monday morning, a serious AI manager should be able to answer:

1. For my top AI systems, can I show a one-page accountability chain with named owners across data, model, deployment, and operations?
2. Do we have clearly defined decision rights for deployment, risk acceptance, and kill switches—and do people know them?
3. Are approval checkpoints and evidence requirements documented and understood by project teams?
4. If an AI-related incident happened today, would my team know whom to call, what to log, and how to escalate within the hour?

5. When did we last run an accountability-focused review or drill for a high-impact AI system, and what changed afterward?

If your answers are mixed, that is your roadmap. Building a robust accountability chain will not only protect your organization when AI fails—it will give your leadership the confidence to let AI scale, knowing someone is genuinely in charge at every link.

Would you like a **combined one-page "AI System Accountability Sheet"** that merges the mapping, approval, and escalation fields into a single template for your governance toolkit?

7 Policy and Compliance in Practice

Opening vignette

You are mid-way through an AI modernization program when your General Counsel forwards an email from a regulator asking three questions: which of your AI systems fall under "high-risk" classifications, how you manage AI risk, and how you ensure consistent treatment of users across regions. Your team has good technical controls, but your answers are scattered in slide decks, emails, and one-off documents. Legal wants clarity. The COO wants reassurance. The board wants proof you are ahead of emerging AI rules, not waiting to be caught. In that moment, you see the gap: you have AI systems and pockets of governance, but you do not yet have policy and compliance built into the way work gets done.

Why AI policy and compliance matter for modernization leaders

AI policy and compliance are often treated as legal background noise. For IT managers leading modernization, these operational guardrails protect your license to innovate. The regulatory environment is tightening:

- The EU AI Act introduces a tiered, risk-based regime with strict obligations for high-risk systems, including risk management, data quality, logging, documentation, human oversight, and robustness requirements that kick in between 2026 and 2027.
- The U.S. is moving toward a national AI policy framework through executive orders and guidance that aim to harmonize state laws and set baseline expectations for risk, fairness, and transparency.
- Frameworks like the NIST AI Risk Management Framework (AI RMF) give voluntary, but influential, guidance on how to govern AI across its lifecycle through functions like Govern, Map, Measure, and Manage.

These are not abstract. They directly shape what "good" looks like when you roll out AI in production.

From Policy to Practice

Global / national AI rules

e.g., EU AI Act, US Executive Orders

Enterprise policies & standards

e.g., Company AI Code of Conduct, internal compliance frameworks

Project workflows & controls

e.g., Risk assessment checklists, data quality checks, deployment controls

7.1 Reframing compliance: from paperwork to risk prevention

Most managers think of compliance as forms, checklists, and "slowing things down." In AI, that mindset is dangerous. The cost of non-compliance is not only fines; it is emergency remediation, reputational damage, and loss of executive confidence.

Reframed correctly, AI policy and compliance help you:

- Identify high-risk systems early and treat them accordingly.
- Avoid surprise regulatory obligations late in the project.
- Standardize controls so you are not reinventing governance for every use case.

As a modernization leader, your task is to embed the **minimum necessary controls** in a way that feels like part of the build process, not a separate compliance track bolted on at the end.

Compliance as Risk Prevention

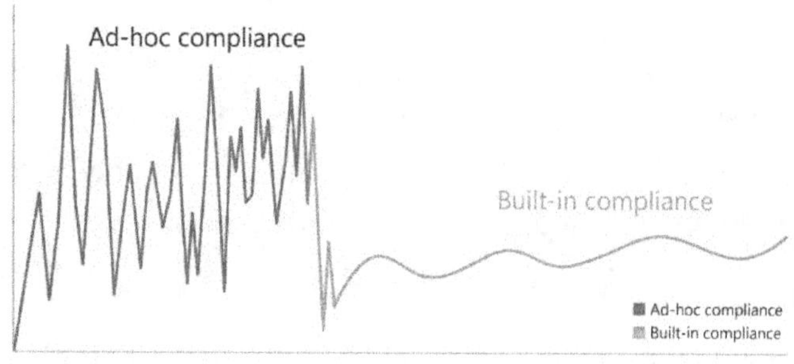

7.2 The evolving AI policy landscape: what you must track

You do not need to become a regulatory expert, but you do need a working map of the AI policy environment your organization operates in:

- **EU AI Act:** Risk-based categories (prohibited, high-risk, limited risk, minimal risk) with strict obligations for high-risk systems, including risk assessment, data quality, logging, documentation, human oversight, and robustness. europarl.
- **U.S. federal direction:** A series of executive orders and policy steps aimed at creating a unified national AI framework, limiting conflicting state laws, and promoting "trustworthy" AI in government and industry.

- **Voluntary frameworks:** NIST's AI RMF, widely referenced by enterprises, which emphasizes four core functions—Govern, Map, Measure, Manage— as a way to organize risk management activities.

Your legal and compliance teams will handle the detailed interpretation. Your job is to understand which systems are in scope, what kinds of controls regulators expect, and how to build those controls into your modernization approach.

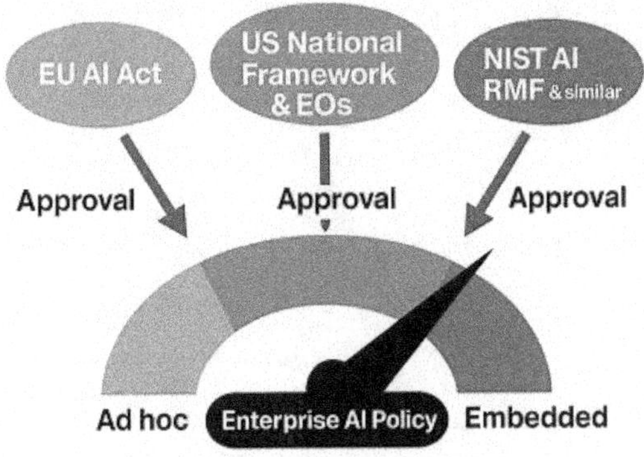

7.3 Translating external rules into internal policies

External rules provide direction; internal policies turn them into concrete obligations. As an IT manager, you should help shape internal AI policies that:

- **Classify AI systems by risk** (for example, high/medium/low impact) based on factors similar to those in the EU AI Act (safety, rights, access to essential services).
- **Define minimum controls** required at each risk level (documentation, testing, oversight, logging, etc.).
- **Reference AI RMF-style functions** so that risk management steps are aligned with a recognized framework (Govern, Map, Measure, Manage).

This translation step moves you from "we should worry about the EU AI Act and executive orders" to "here is our internal AI policy playbook, and here is what every project must do."

External Rules → Internal Policy → Project Controls

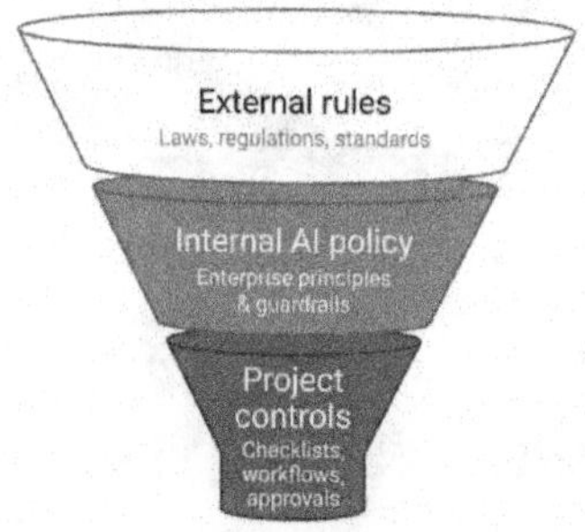

7.4 The manager's AI policy checklist

Use this checklist when you start or review an AI initiative to align it with organizational policy and external expectations:

AI Policy Alignment Checklist

- Scope
  - We know which jurisdictions (EU / US federal / others) are relevant for this system's users and data flows.
  - We have confirmed whether this system might fit a "high-risk" category under applicable rules (e.g., impacts access to services, employment, credit, safety).
- Internal policy fit
  - The system's risk tier (high/medium/low) is assigned per internal AI policy.
  - Required controls for that tier are identified (documentation, testing, oversight).
- Roles and governance
  - Legal/compliance is engaged at the right checkpoints, not just at the end.
  - A governance forum is identified where this system will be reviewed.
- Documentation and evidence
  - We know which documents we must produce (model card, AI fact sheet, risk assessment, DPIA where relevant).
  - We know how we will store and retrieve these artifacts for audits.

7.5 The EU AI Act risk-based approach: practical implications

The EU AI Act is the most detailed, risk-based AI regulation to date. Even if your organization is not EU-based, it is widely shaping expectations of "good practice." In simplified terms: digital strategy.

- **Prohibited AI:** Certain uses (e.g., specific manipulative or social scoring systems) are banned. You should treat similar patterns with extreme caution everywhere.
- **High-risk AI:** Systems used in domains like credit, employment, education, essential services, critical infrastructure, and certain biometrics are subject to strict obligations (risk management, data quality, logging, documentation, human oversight, robustness, registration).
- **Limited-risk AI:** Some systems must meet transparency obligations (for example, letting users know they are interacting with AI).
- **Minimal-risk AI:** Many internal productivity tools may fall here with minimal explicit obligations.

For modernization leaders, the key move is to align your risk classification and control expectations with this style of thinking: high-impact systems get high-rigor controls.

Risk–Tiered Controls (EU–Inspired)

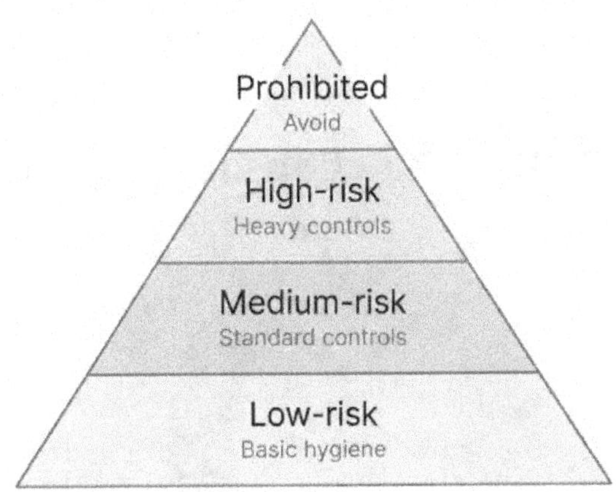

NIST AI RMF as an organizing backbone

The NIST AI RMF is voluntary, but it is fast becoming a common reference for structuring AI governance. It organizes activities into four core functions:

- **Govern:** Establish AI risk governance, roles, policies, and culture.
- **Map:** Understand context, use case, stakeholders, and potential harms.
- **Measure:** Assess risk, performance, fairness, and other trustworthiness characteristics.
- **Manage:** Prioritize and implement risk controls, monitor, and respond.

As an IT manager, you can use these as a mental model: every AI project should show how it is being governed, mapped, measured, and managed—not as abstract language, but as concrete steps and artifacts.

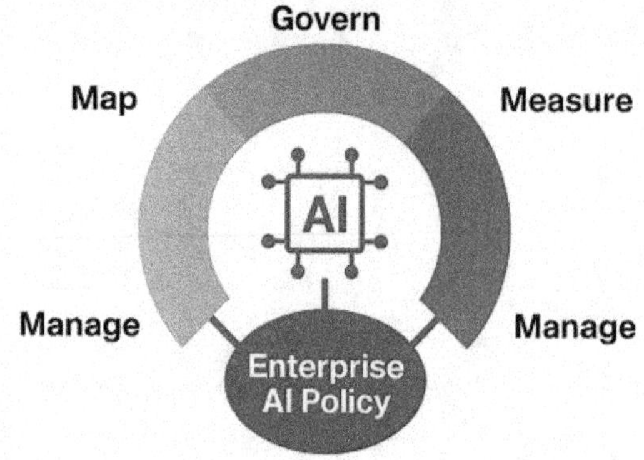

Template: AI policy control matrix (by risk tier)

This matrix turns policy principles into concrete requirements. Customize the details for your organization.

AI Policy Control Matrix – Template

Columns:

- Control area (Use case, Data, Model, Fairness, Safety, Transparency, Logging, Oversight, Incident response).

- Low-risk systems: minimum expectations.
- Medium-risk systems: standard expectations.
- High-risk systems: enhanced expectations.

Example rows (to be tailored):

- Use case approval
    - Low: Manager approval.
    - Medium: Governance review.
    - High: Formal risk assessment + senior sign-off.
- Data documentation
    - Low: Basic description.
    - Medium: Source, time range, basic assessment.
    - High: Detailed data governance and quality assessment.
- Fairness and bias
    - Low: Basic consideration.
    - Medium: Some testing on key segments.
    - High: Structured bias audit and monitoring plan.

Policy Control Matrix Layout

	Low–risk	Medium–risk	High–risk
Data & access			
Model & testing			
Monitoring & incidents			
Documentation			

7.6 Building compliance with workflows

Once you have defined controls, the critical move is integration: your modernization workflows (intake, design, build, deploy, operate) should **trigger the right compliance steps by default**, rather than relying on people to remember them.

Examples of embedding compliance:

- Intake forms that capture risk tier, jurisdictions, and key policy triggers.
- Design review templates that include AI policy questions (e.g., high-risk use case, cross-border data flows).
- Definition of Done (DoD) checklists for sprints that include documentation and evidence tasks.

- Change management workflows require updated risk assessments for high-impact changes.

Your goal is for teams to experience compliance as "how we do things," not "extra work we bolt on at the end."

Embedded Compliance Workflow

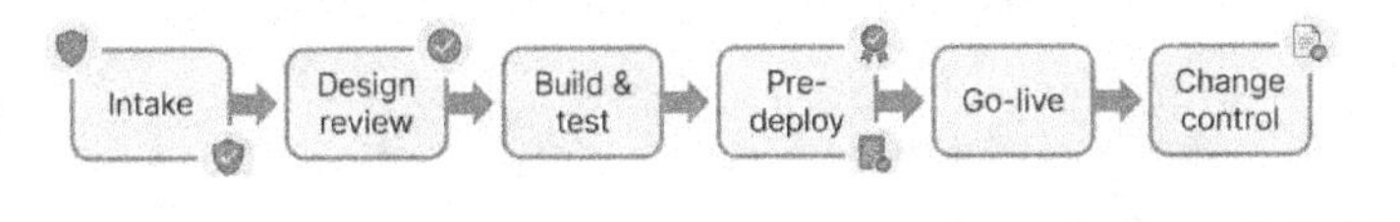

7.7 Template: AI project intake & policy trigger form

Use this form (or its fields) for every new AI initiative.

AI Project Intake & Policy Trigger – Template

1. Basic information

- Project name:
- Business area:
- Project owner:

2. Use case and impact

- Brief description of AI use:
- Types of decisions affected (advisory, assistive, automated):

- Impacted populations (customers, citizens, employees):
- Does this affect access to services, employment, credit, health, or safety? [Yes/No]

3. Jurisdictions and data

- Countries/regions of users:
- Locations of data storage and processing:
- Presence of EU users or data: [Yes/No]

4. Preliminary risk tier (per internal policy)

- Proposed risk tier: [Low / Medium / High]
- Rationale (2–3 sentences):

5. Policy triggers

- Possible EU AI Act "high-risk" category? [Yes/No/Uncertain]
- Involves sensitive attributes (health, biometrics, protected characteristics)? [Yes/No]
- Requires explicit fairness assessment. [Yes/No]
- Requires DPIA or similar impact assessment? [Yes/No]

6. Governance & approvals

- Proposed governance forum for review:
- Legal/compliance contact:

7.8 Compliance evidence: what you will be asked to show

When regulators, auditors, or internal assurance teams come knocking, they will not just ask what your policy says; they will ask what you can **show**. Common categories of evidence include:

- Use case assessments and risk classifications for high-impact systems.
- Documentation: model cards, AI fact sheets, data documentation, risk assessments.
- Records of testing (performance, fairness, safety) and their results.
- Governance minutes: evidence of reviews, approvals, and risk decisions.
- Operational logs: decision logs, incident reports, and remediation actions.

As an IT manager, you should design your processes so that these artifacts are created as a natural byproduct of doing the work, not as a scramble when someone asks.

Evidence Pyramid

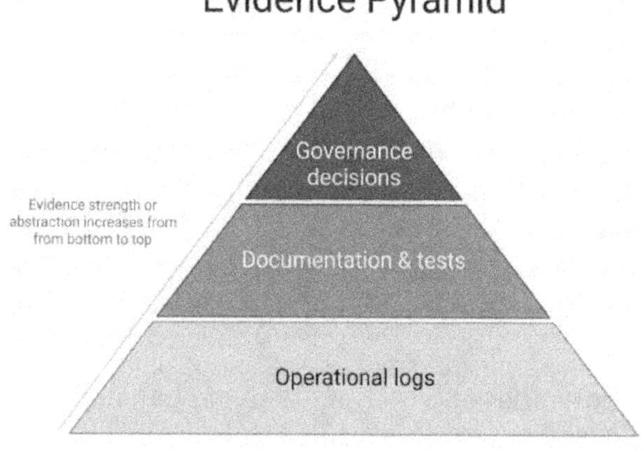

Manager checklist: AI compliance readiness (system level)

For each significant AI system, use this checklist to gauge readiness:

AI Compliance Readiness Checklist

- Risk classification
    - Risk tier assigned and documented.
    - Rationale aligns with internal policy and external expectations.
- Policy controls
    - Required controls (for this tier) are implemented and documented.
    - Outstanding control gaps are known and managed (with risk acceptance if needed).
- Documentation

- o Model card and AI fact sheet were completed.
  - o Data sources and governance documented.
- Governance
  - o System has been reviewed in the appropriate governance forum.
  - o Approvals (or conditions) are recorded.
- Operations
  - o Logging, monitoring, and incident processes are in place.
  - o Compliance-relevant events (incidents, complaints) are tracked.

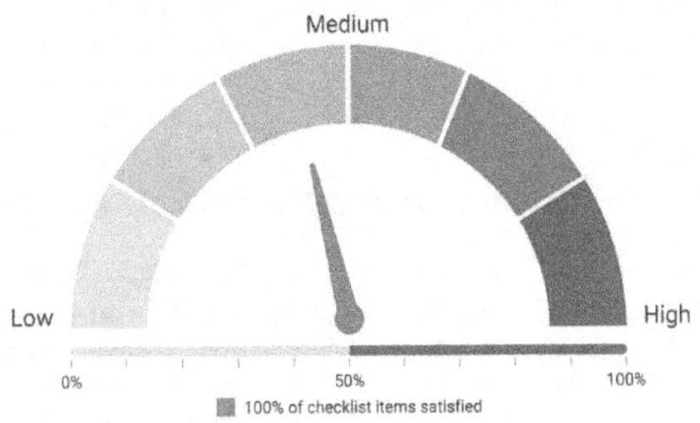

Compliance Readiness Dial

Case pattern: EU-exposed high-risk system

Imagine you are deploying an AI system to support loan approvals for a business serving EU customers. Under the

EU AI Act, credit-related systems are likely to be treated as high-risk, triggering obligations such as risk management, data quality, documentation, logging, human oversight, and robustness.

Practically, for you, this means:

- You must classify the system as high-risk internally and implement higher controls.
- The system may need registration and documented conformity steps (even if handled by another corporate function).
- You must be able to show robust documentation and log to European regulators if questions arise.

Treating this like any other system is not an option; policy and compliance must shape the architecture and roll out from day one.

High-Risk EU System Controls Map

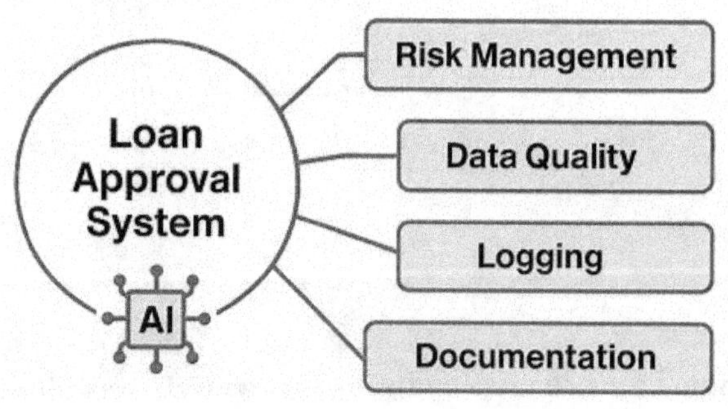

Case pattern: U.S.-centric system in a harmonizing landscape

Now consider an AI system used primarily in the U.S., where federal executive orders and national frameworks aim to harmonize state rules while promoting innovation. The formal obligations may be less prescriptive than the EU AI Act, but expectations around fairness, transparency, and "trustworthy AI" are growing.

For IT managers, this environment means:

- You cannot just optimize for the least restrictive state or sector; national guidance is pushing toward consistent baseline controls.
- You should align with federal "trustworthy AI" principles and NIST AI RMF to demonstrate diligence.
- You must anticipate evolving expectations and avoid locking into architectures that would be expensive to retrofit with controls later.

Here, compliance is about futureproofing: building enough discipline now so you do not face painful, large-scale rework as the U.S. regime tightens.

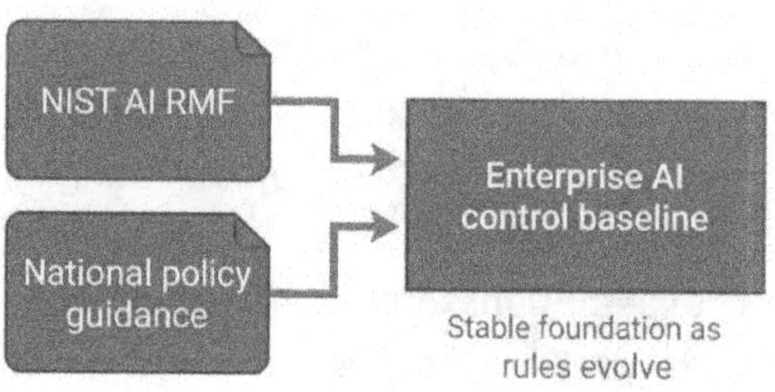

The Manager's Policy & Compliance Playbook (120-day plan)

Policy and compliance are broad; you need a focused, staged plan.

Days 0–30: Establish the baseline

- Meet with legal/compliance to understand key external AI obligations relevant to your portfolio (EU, US, others).
- Inventory AI systems and assign a preliminary risk tier to each.
- Identify which systems likely fall into "high-risk" categories under external or internal rules.

Days 31–60: Define and codify controls

- Work with risk/compliance to finalize your AI policy control matrix by risk tier.

- Embed the AI project intake & policy trigger form into your demand/intake process.
- Identify the governance forums where high-risk systems will be reviewed.

Days 61–90: Implement priority systems

- For your top 3–5 high-impact systems, ensure control requirements are implemented or planned.
- Complete or refresh documentation (model cards, fact sheets, risk assessments).
- Define how evidence (logs, minutes, approvals) is stored and retrieved.

Days 91–120: Test and refine

- Run a compliance "tabletop" exercise: simulate an AI audit or regulatory inquiry for one high-risk system.
- Document gaps found and feed them back into your control matrix and workflows.
- Adjust templates and checklists based on real-world feedback from project teams.

120-Day Policy Implementation Roadmap

Days 0–30	Days 31–60	Days 61–90	Days 91–120
Define scope & objectives	Draft AI policy & standards	Pilot controls & checklists	Roll out policy
Identify high-risk use cases	Consult key stakeholders	Set approval workflows	Train teams
Assemble cross-functional team	Obtain legal & compliance review	Integrate tools & technology	Monitor & refine
Conduct gap analysis	Develop communication plan	Assess pilot results & adjust	

7.9 The AI Reality Check: avoiding "policy theater."

"Policy theater" happens when organizations publish AI principles, adopt frameworks by name, and add a few new slides—without changing decisions. As an IT manager, you know you are in policy theater if:

- Risk tiers are defined, but projects are not actually classified.
- Policies exist, but no one can show how they map concrete controls.
- Governance forums meet, but few decisions or actions are recorded.
- Evidence is scattered and hard to retrieve when asked for.

Real policy and compliance work will change how you prioritize projects, design systems, and manage change. If nothing about your delivery process changes, your policy is a poster, not a practice.

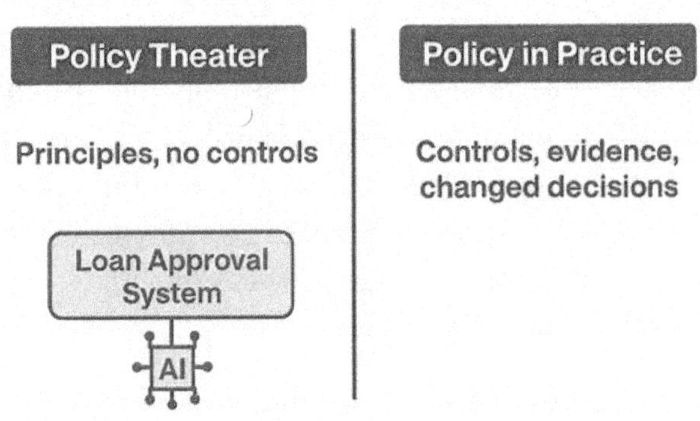

The Monday Morning Test

On Monday morning, a serious AI manager should be able to answer:

1. For each major AI system I own, what is its risk tier and why?
2. Which external AI rules (EU AI Act, U.S. national framework, others) likely apply to it, directly or indirectly?
3. What specific policy-driven controls are implemented for this system (documentation, testing, oversight, logging, incident response)?

4. If a regulator or internal auditor requested evidence tomorrow, could we produce the key artifacts within days, not weeks?
5. When did we last review our AI policy control matrix or run a policy-focused drill, and what changed as a result?

If some answers are "not yet," that is not a failure; it is your roadmap. Policy and compliance in practice are not about creating fear; they are about giving your AI modernization program a stable foundation so you can move fast where it is safe—and slow where it matters.

Policy & Compliance Maturity Gauge

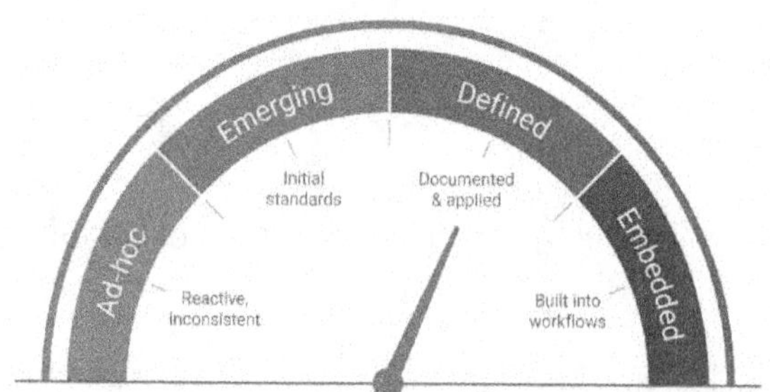

8 Managing AI Risk: Frameworks That Work

Opening vignette

You are leading an AI modernization program with half a dozen models already live: a routing model in customer service, a fraud model in finance, a triage model in IT operations, and a generative assistant for knowledge workers. Individually, each project has done its own risk assessment. Collectively, you sense something is missing: there is no shared language for risk, no consistent scoring, no way to compare a hallucinating chatbot against a mis-prioritizing fraud model. In a steering committee meeting, the CFO asks: "Which AI risks should we be most worried about this quarter—and what are we doing about them?" Your team looks at one another, each thinking in their own silo.

In that moment, you realize you do not just need controls; you need **frameworks** that tie risks together across systems, teams, and domains.

8.1 Why risk frameworks sit at the heart of operational governance

AI risk is different from traditional IT risk. It is not just outages and breaches; it is also biased decisions, unsafe content, misuse, and a strategic reliance on models you do not fully control. Without a structured framework, risk management becomes reactive and anecdotal: people raise isolated issues, but you lack a coherent view of where risk originates, how big it is, and what trade-offs you are making.

Risk frameworks do three things for modernization leaders:

- Provide a **shared vocabulary** for different stakeholders (IT, business, legal, risk) to talk about AI risk.
- Ensure risks are **evaluated consistently** across use cases, not based on who shouts the loudest.
- Create a **repeatable process** to prioritize mitigations and balance innovation with control.

From Ad-hoc Concerns to Structured Risk

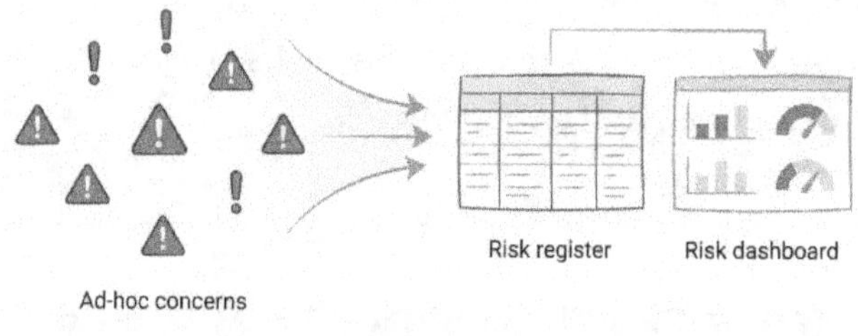

8.2 The core idea: value, feasibility, and impact

At the manager level, three axes help you reason about AI risk in a practical way:

- **Value:** What is the business or mission value of this AI system (efficiency, quality, reach, revenue, public impact)?
- **Feasibility:** How feasible is it to reduce or control the risk (availability of mitigations, technical complexity, resource needs)?
- **Impact:** If things go wrong, how serious are the consequences (harm to people, regulatory exposure, reputational damage, operational disruption)?

You want to **focus on heavy controls where impact is high, and mitigation is feasible**, and be explicit when you accept risk because value is high and mitigation is limited.

Value–Feasibility–Impact Triangle

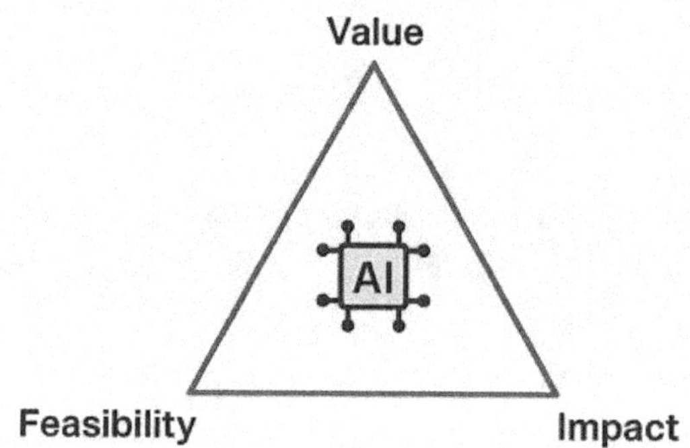

Introducing NIST's AI Risk Management Framework (AI RMF)

The NIST AI RMF provides a widely recognized structure for managing AI risk, built around four core functions: **Govern, Map, Measure, and Manage**.

- **Govern:** Establishes roles, responsibilities, policies, and a risk-aware culture.
- **Map:** Understands the context, use case, stakeholders, and potential harms.
- **Measure:** Assesses risks using metrics, testing, and evaluation methods.
- **Manage:** Prioritizes and treats risks and implements monitoring and response.

For IT managers, this is not theory; it is a way to organize your AI governance activities to ensure consistency across projects.

NIST-Style AI RMF Functions

8.3 Introducing ISO 42001: AI management systems

ISO/IEC 42001 is the emerging standard for AI management systems. It is like ISO 27001 for cybersecurity but focused on AI: a set of requirements and guidance for establishing, implementing, maintaining, and improving an AI management system.

Key elements relevant to modernization leaders include:

- **Governance structures** for oversight and accountability.

- **Risk management protocols** to identify, assess, and mitigate AI risks across the lifecycle.
- **Responsible AI practices** covering fairness, transparency, and reliability.
- **Continuous improvement** of your AI processes and controls.

You do not need to be ISO-certified to benefit. You can treat ISO 42001 as a blueprint for how your AI governance operating model should look and feel.

ISO 42001 AI Management System

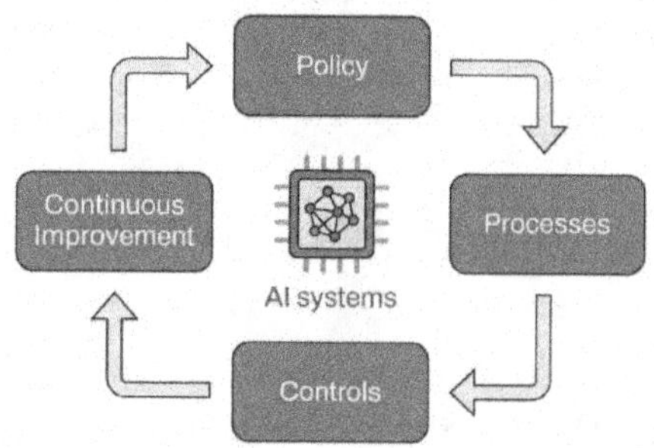

8.4 Making frameworks usable: a manager's translation layer

NIST AI RMF and ISO 42001 can feel abstract to delivery teams. Your job is to translate them into **practical tools**:

checklists, templates, and scoring methods that show up in real workflows.

A simple translation strategy:

- Use **NIST's functions** (Govern, Map, Measure, Manage) as the backbone of your AI project lifecycle.
- Use **ISO 42001** as a reference when defining your organization-wide AI governance processes and control catalog.
- Add **value–feasibility–impact scoring** on top to prioritize actions across your portfolio.

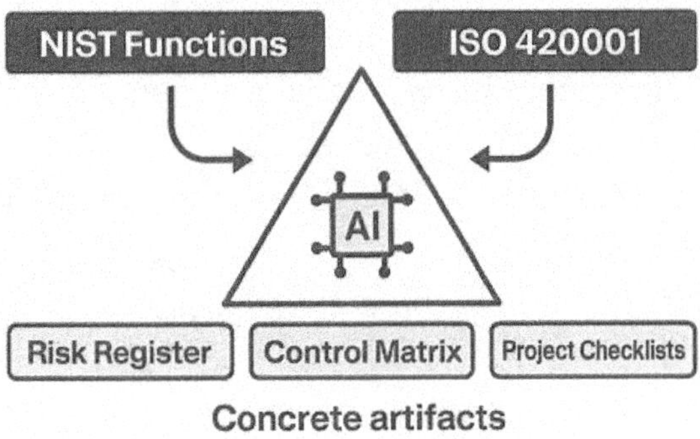

Template: AI risk register (system level)

A risk register is your central tool for tracking AI risks. Here is a manager-friendly template:

AI Risk Register – Template (Per System)

For each identified risk, record:

- Risk ID:
- System name:
- Risk description (short, concrete):
- Category (e.g., Safety, Fairness, Security/Misuse, Compliance, Reputation, Operational):
- Root cause (data, model, workflow, user behavior, vendor, etc.):
- Likelihood (1–5):
- Impact (1–5):
- Value of system (Low/Med/High) for context:
- Feasibility of mitigation (1–5):
- Current controls in place:
- Proposed additional controls:
- Residual risk after mitigation (Low/Med/High):
- Owner (name/role):
- Status (Open / In progress / Closed):

Risk Register Layout

Risk ID	Category	Likelihood	Impact	Controls	Owner

Consistent scoring: likelihood and impact

To compare risks across systems, you need a **consistent scoring scheme**. A simple 1–5 scale for both likelihood and impact works well.

Example impact scale:

- 1 – Negligible: Minimal effect, easily reversible, small scope.
- 2 – Minor: Limited issues for a small group; no regulatory or media interest.
- 3 – Moderate: Noticeable impact on operations or user experience; manageable but visible.
- 4 – Major: Significant harm to individuals or operations; potential regulatory or media attention.
- 5 – Severe: Serious harm, systemic disruption, high regulatory and reputational impact.

Example likelihood scale:

- 1 – Rare: Unlikely, but not impossible.
- 2 – Unlikely: Could happen under unusual conditions.
- 3 – Possible: Could reasonably occur during normal operations.
- 4 – Likely: Expected to occur periodically.
- 5 – Almost certain: Expected to occur frequently without additional controls.

Multiply or combine these to derive a **risk rating** (e.g., Low/Medium/High based on thresholds).

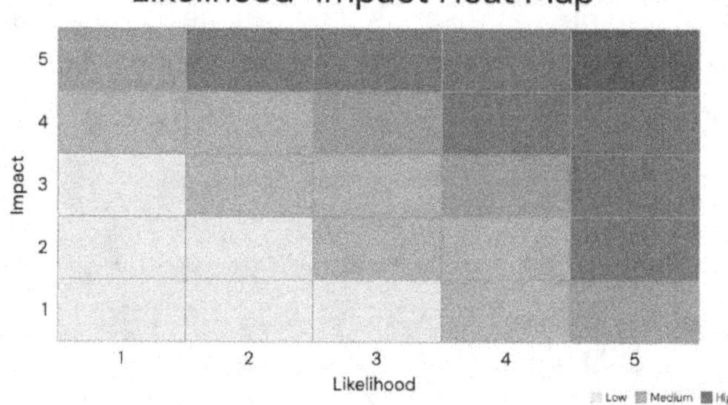

Layering in value and feasibility

Not all high-scoring risks are equal. A high-impact risk in a low-value experimental tool may warrant a different treatment than a similar risk in a high-value, core system. Similarly, if mitigation is extremely difficult or costly, you may choose targeted controls and strongly governed risk acceptance instead of full mitigation.

To make this explicit, add:

- **Value rating** (Low/Med/High) – how central the system is to your strategy.
- **Feasibility score** (1–5) – how realistic it is to reduce this risk significantly.

This supports decisions like:

- "High risk, high value, high feasibility → invest in strong controls."
- "High risk, low value → consider avoiding or redesigning the use case."
- "High risk, high value, low feasibility → tighten oversight and document risk acceptance."

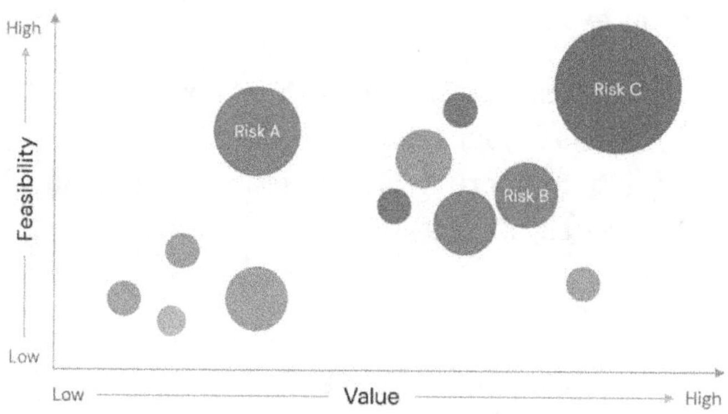

8.5 Domain adaptations: public sector, healthcare, finance

Risk frameworks are generic; domains are not. AI risk plays out differently in different environments:

- **Public sector:** High sensitivity to fairness, due process, and rights. AI decisions around benefits, enforcement, or public services carry heightened scrutiny and political risk. Frameworks like NIST AI RMF align well with public-sector risk and human rights considerations.

- **Healthcare:** Patient safety, clinical validity, and data privacy dominate. AI risk management must integrate with existing clinical governance and safety frameworks (e.g., tying NIST AI RMF into COSO ERM and clinical review processes).

- **Finance:** Fraud, credit, AML, and market integrity are heavily regulated. AI risk must align with existing enterprise risk management, model risk management, and regulatory expectations around fairness, explainability, and stress testing.

Your job is to align generic AI frameworks with **domain-specific governance** you likely already have.

Domain-Specific Risk Adaptation

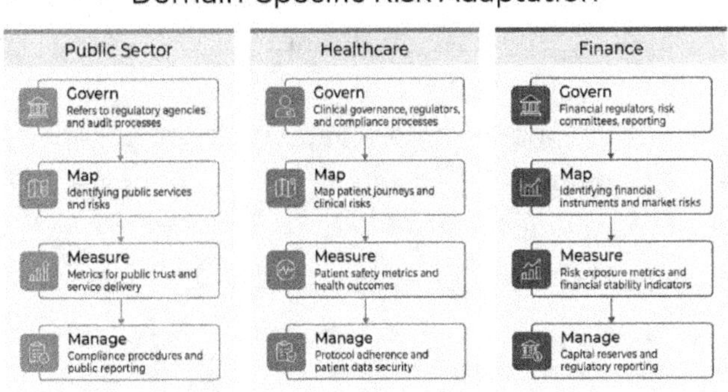

Template: AI risk assessment worksheet (per use case)

Before building or deploying, use this structured worksheet:

AI Risk Assessment Worksheet – Template

1. Context (Map / Govern)

- System name and description:
- Business process and objectives:
- Stakeholders and affected groups:
- Domain (public sector/healthcare/finance / other):

2.   Risk   identification   (Map)

For each risk category, list key risks:

- Safety (harmful outputs, unsafe recommendations):
- Fairness (biased outcomes, disparate treatment):
- Security/misuse (prompt injection, data exfiltration, abuse):
- Privacy (inappropriate data use or leakage):
- Compliance (regulatory violations, policy breaches):
- Reputation (public trust, media exposure):
- Operational (downtime, brittle behavior, dependencies):

3.   Scoring   (Measure)

For each risk:

- Likelihood (1–5):
- Impact (1–5):
- Value of system (Low/Med/High):
- Feasibility of mitigation (1–5):
- Risk rating (derived):

4.   **Treatment**   **(Manage)**

For each high or medium risk:

- Proposed controls:
- Residual risk estimate:
- Owner:
- Target timeline:

Manager checklist: AI risk governance baseline

This checklist helps you confirm whether AI risk management is systematic rather than ad-hoc:

AI Risk Governance Baseline Checklist

- Framework usage
    - We have chosen reference frameworks (e.g., NIST AI RMF, ISO 42001) and use their language in our internal processes.
    - Our AI governance policies reference these frameworks explicitly.
- Risk processes
    - Each significant AI system has a documented risk assessment using a standard template.
    - We maintain an AI risk register across systems, not just within projects.
- Scoring and prioritization

- We use consistent scales for likelihood and impact across AI projects.
  - We consider system value and feasibility of mitigation when prioritizing actions.
- Domain fit
  - Our risk approach is tuned to specific domains (e.g., public sector, healthcare, finance) where we operate.
  - Domain experts participate in risk discussions, not just IT and data teams.

8.6 Portfolio-level risk: seeing across systems

Managing risk one system at a time is not enough once your AI footprint grows. You need a **portfolio view**:

- Which AI systems carry the highest residual risk?
- Where do multiple systems share the same root causes (e.g., same data source, same vendor, same pattern of misuse)?
- How is AI risk trending over time (more systems, more incidents, improving controls)?

A simple AI risk portfolio dashboard might track:

- Number of AI systems by risk tier.
- Top 10 residual risks with owners and status.
- Incidents by category and month.

- Progress on key mitigation initiatives.

Template: AI risk portfolio summary (quarterly)

Every quarter, produce a management-level summary:

8.7 AI Risk Portfolio Summary – Template

- Period covered:
- Total number of AI systems in scope:
- Distribution by risk tier: [Low/Med/High counts]
- Top 5 systems by residual risk (name, domain, brief description).
- Top 10 portfolio-level risks (risk ID, description, owner, status).
- Notable incidents and lessons learned this period.
- Key risk reduction initiatives underway (with target dates).
- Assessment vs previous quarter (risk increased/decreased, key drivers).

This becomes your primary tool for answering "What AI risks matter most right now?"

Bringing it together: Govern–Map–Measure–Manage in your workflow

To avoid frameworks becoming shelfware, embed the NIST-style cycle directly into your AI delivery workflow:

- **Govern:** Define roles, policies, risk appetite; align with ISO 42001-style management system. learn.
- **Map:** Use the AI risk assessment worksheet early in each project.
- **Measure:** Apply consistent scoring, test methods, and monitoring metrics.
- **Manage:** Maintain the risk register, implement controls, and track portfolio status.

The point is not perfection; it is **repeatability**. Every system follows the same spine, adjusted for its risk tier and domain.

Govern–Map–Measure–Manage Embedded in Lifecycle

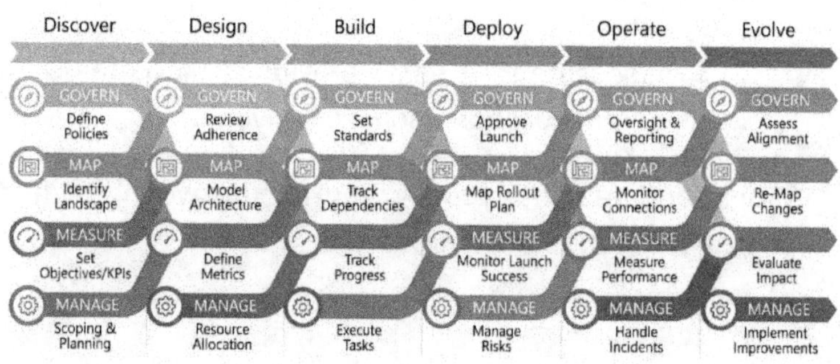

8.8 The Manager's AI Risk Playbook (90-day plan)

To operate risk frameworks in your modernization program, use this 90-day playbook:

0–30 days: Choose and align

- Align with risk/compliance on using NIST AI RMF as your project-level structure and ISO 42001 as a management-system reference.
- Define standard scales for likelihood and impact.
- Agree on value and feasibility as decision lenses.

31–60 days: Implement for priority systems

- For the top 3–5 AI systems, complete the AI risk assessment worksheet.
- Create or update risk registers for each and a simple portfolio summary.
- Link risk owners and mitigation actions to your governance forums.

61–90 days: Institutionalize

- Embed the risk assessment worksheet into AI project intake and design reviews.
- Establish a quarterly AI risk portfolio review with leadership.
- Tune the approach for key domains (public sector, healthcare, finance) with domain experts.

AI Risk Playbook Roadmap

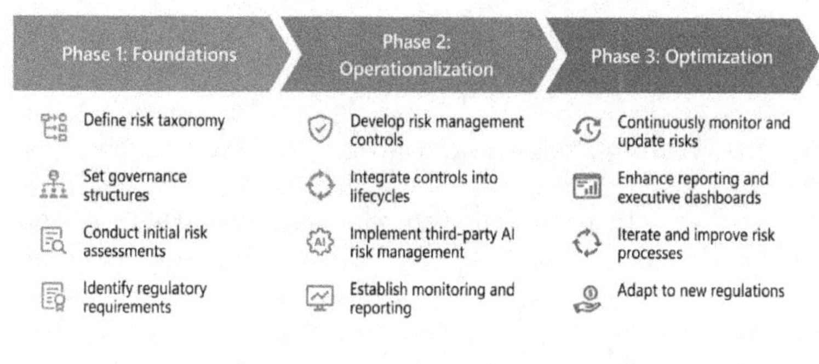

8.9 The AI Reality Check: avoiding "framework theater."

Just as there is ethics theater and policy theater, there is **framework theater**: adopting big names like NIST AI RMF and ISO 42001 in slide decks without changing how risks are actually handled. You know you have slipped into framework theater if:

- Teams can quote the framework functions, but cannot show a completed risk assessment.
- There is no shared risk register or portfolio view.
- Risk ratings feel arbitrary and vary wildly between projects.

- Frameworks are referenced in documents but not in decisions (e.g., priority calls, go/no-go, control investments).

Real use of frameworks shows up in **how you prioritize work** and **how you respond to issues**—not just in how you talk.

Framework Theater vs Framework in Practice

Framework Theater	Framework in Practice
slogans	decision rights
posters	workflows
buzzwords	artifacts
one-off workshops	metrics
	recurring forums

The Monday Morning Test

On Monday morning, a serious AI manager should be able to answer:

1. Which AI risk framework(s) are we actually using (in more than just name), and where do they show up in our project templates and reviews?
2. For my top AI systems, can I open a risk register that shows identified risks, scores, owners, and status?

3. How do we consistently score AI risks across projects, and who agreed to that scoring scheme?
4. What does our AI risk portfolio summary say are the top 5 risks this quarter—and what are we doing about each?
5. How have our risk frameworks influenced a real decision in the last 90 days (e.g., delaying deployment, adding controls, or scoping a use case)?

If some answers are "not yet," you now have a concrete roadmap. Frameworks are not about adding complexity; they are about giving your AI modernization program a **common operating picture** of risk so you can move fast when it is safe—and slow, deliberately, when it is not.

AI Risk Maturity Gauge

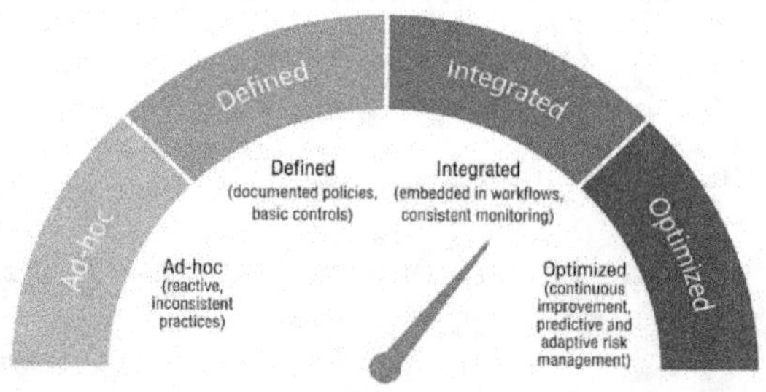

9 The Governance Operating Model

Opening vignette

You have got more AI in production than ever before: a few models built in-house, several from cloud vendors, plus a generative assistant pilot. Every month, a new team comes up with a new idea. You have set up some review meetings, created a few policies, and asked teams to "loop you in." However, issues keep popping up: an AI chatbot used outside its approved scope, a model pushed to production without risk review, a vendor model updated silently overnight. In a steering meeting, your COO asks: "What is our AI governance model? Who approves what, and how do we know it is working?" You realize you do not just need more controls—you need a **governance operating model** that ties everything together.

9.1 Why an AI governance operating model matters

AI governance is not a single committee or policy document. It is a **system**: the roles, processes, oversight bodies, and metrics that guide how AI is proposed, built, deployed, and monitored. Without a clear operating model:

- Decisions are inconsistent across teams.
- Risk is managed unevenly—too strict in some areas, too loose in others.
- Governance feels like surprise gates rather than clear pathways.

A well-designed governance operating model does three things for IT managers:

- Makes it clear **who does what**, when, and with what authority.
- Embeds governance **into workflows**, rather than layering it on top.
- Provides **feedback loops** so you can see whether governance is improving outcomes.

What a Governance Operating Model Covers

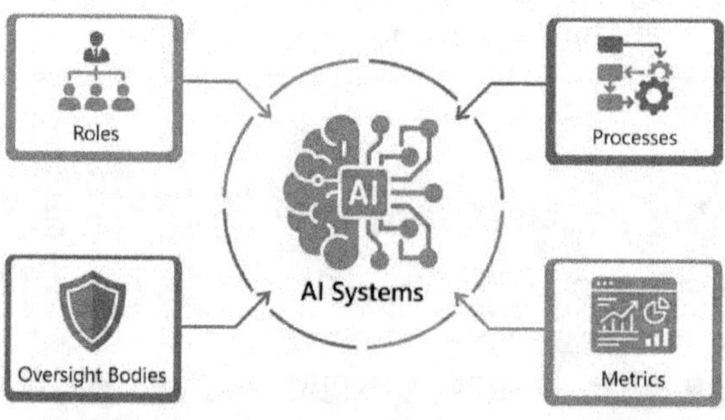

9.2 Governance as workflow, not bureaucracy

Many teams fear governance because they associate it with slow approvals and red tape. The shift you need to lead is to present governance as a **guided workflow**:

- Clear entry points (how ideas come in).
- Predictable steps (what evidence and reviews are required).
- Defined exits (how to get to "go live," "re-scope," or "no go").

When governance is a workflow, teams know what to expect and how to prepare. When it is a collection of ad hoc meetings, everything feels like a one-off exception.

From Gates to Guided Flow

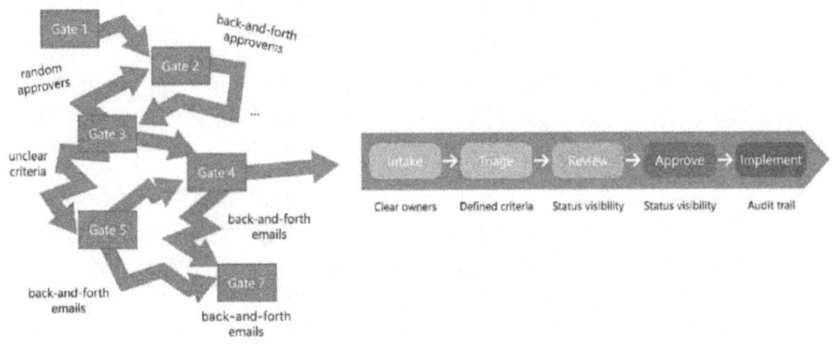

9.3 The "AI Governance Operating Model" framework

At a practical level, your AI governance operating model can be framed around four components:

1. **Structure:** The roles and oversight bodies that make and review decisions.
2. **Flow:** The end-to-end process from idea to retirement.
3. **Controls:** The required artifacts, checks, and approvals at each stage.
4. **Signals:** The metrics and feedback loops that show whether governance is working.

As an IT modernization leader, your task is to define each of these in a way that fits your organization's size, culture, and risk profile.

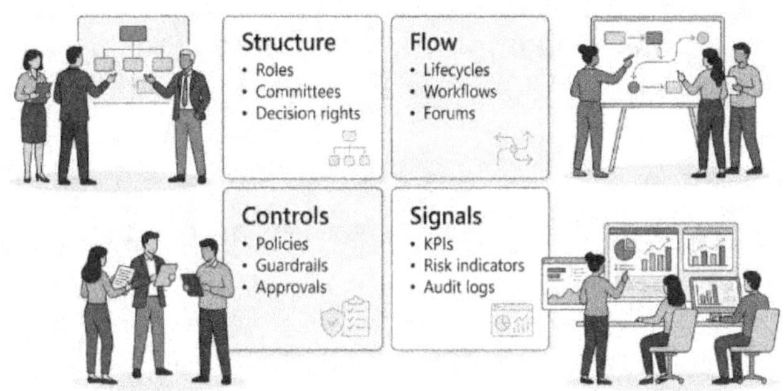

Structure: defining roles and oversight bodies

Start with structure: **who** is involved in AI governance and **at what level.** Typical elements include:

- **AI Steering Committee / Council:** Senior group that sets direction, risk appetite, and resolves major escalations.
- **AI Review Board / Working Group:** Cross-functional body (IT, data, legal, risk, business, HR) that reviews high-impact use cases and major changes.
- **Domain or Business Owners:** Accountable for AI in their areas (e.g., claims, lending, citizen services).

- **Model Owners / Product Owners:** Responsible for specific AI systems and their documentation, performance, and risk.
- **Risk & Compliance Partners:** Provide challenge and assurance at key points.

You do not need dozens of committees. You need **clear, named bodies** with defined scopes and decision rights.

Governance Structure Stack

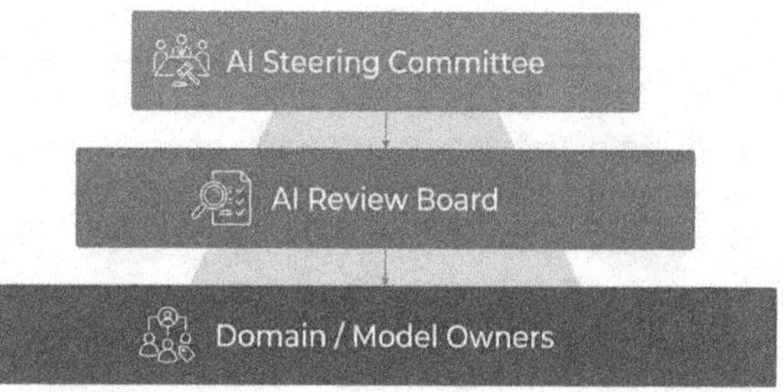

Template: AI governance charter (for a review board)

Use this template to define your AI Review Board:

AI Review Board Charter – Template

1. Purpose

- Provide structured review and challenge for high-impact AI systems.

- Ensure alignment with AI policy, risk appetite, and strategic goals.

2. Scope

- New AI use cases above the defined risk tier.
- Major changes to existing high-impact AI systems.
- Significant incidents or escalations involving AI.

3. Membership

- Chair: [Role/title]
- Core members: [IT, Data/ML, Business, Legal, Risk/Compliance, Security, HR (if relevant)]
- Ad hoc participants: [Domain experts, vendor reps, etc.]

4. Responsibilities

- Review AI project submissions and supporting documentation.
- Approve, conditionally approve, or reject proposals.
- Request additional testing or controls where needed.
- Escalate high-stakes decisions to the AI Steering Committee.

5. Operating cadence

- Meeting frequency: [e.g., bi-weekly]

- Submission deadlines and required materials: [e.g., model card, risk assessment, fact sheet]
- Decision recording and communication process:

AI Review Board Charter Layout

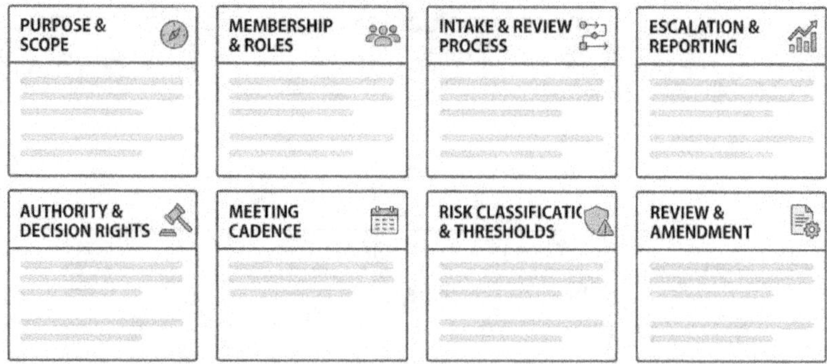

9.4 Flow: the AI governance lifecycle

Your operating model needs a standard **governance flow** for each AI project—with variations based on risk. A simple, effective pattern:

1. **Intake & Triage:** Capture the idea, basic use case, and risk signals.
2. **Assessment & Design:** Perform risk assessment, decide on controls, and design the system accordingly.

3. **Review & Approval:** Present to the appropriate governance body, get decisions and conditions.
4. **Deployment & Monitoring:** Go live under defined controls and monitoring.
5. **Change & Retirement:** Handle updates, expansions, and eventual decommissioning.

High-risk systems may trigger more steps and more intense reviews; low-risk systems move faster with a lighter touch.

AI Governance Flow

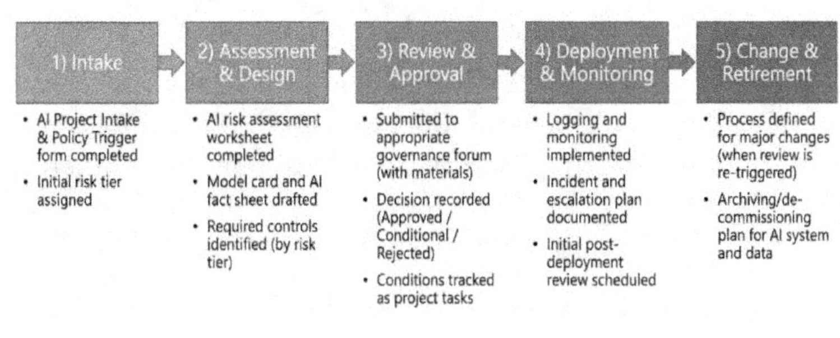

Template: AI governance workflow (project view)

For project teams, you can present this as a simple checklist:

9.5 AI Governance Workflow – Project Checklist

*   Intake

- o AI Project Intake & Policy Trigger form completed.
  - o Initial risk tier assigned.
- Assessment & design
  - o AI risk assessment worksheet completed.
  - o Model card and AI fact sheet drafted.
  - o Required controls identified (by risk tier).
- Review & approval
  - o Submitted to the appropriate governance forum (with materials).
  - o Decision recorded (Approved / Conditional / Rejected).
  - o Conditions tracked as project tasks.
- Deployment & monitoring
  - o Logging and monitoring implemented.
  - o Incident and escalation plan documented.
  - o Initial post-deployment review scheduled.
- Change & retirement
  - o Process defined for major changes (when review is re-triggered).
  - o Archiving/decommissioning plan for the AI system and its data.

9.6 Controls: calibrating governance to risk and culture

Controls are the specific **requirements** a project must meet at each stage: documentation, tests, approvals, and safeguards. The art is calibrating these controls to your:

- **Risk tiers** (low/medium/high).
- **Culture** (centralized vs federated, high-regulation vs more flexible environments).

Examples:

- Low-risk internal productivity tools might need only a short intake, basic documentation, and simple logging.
- High-risk systems (e.g., affecting credit, healthcare, benefits, or safety) might require full risk assessment, bias audits, safety analysis, formal approvals, and frequent reviews.

Your operating model should spell this out so teams do not have to guess.

Template: AI governance control sheet (per system)

Use this template as a one-page summary of the controls applied to a given system:

AI Governance Control Sheet – Template

- System name:
- Risk tier: [Low / Medium / High]

- Business owner:
- Technical owner:

1. Required controls (per policy)

- Documentation: [Model card / Fact sheet / Data documentation]
- Risk assessment: [Y/N, date completed]
- Fairness/bias assessment: [Y/N, scope]
- Safety & misuse assessment: [Y/N, scope]
- Security review: [Y/N, scope]

2. Approvals & reviews

- Governance forum: [Name]
- Last review date & outcome:
- Next scheduled review:

3. Monitoring & incidents

- Key metrics monitored:
- Incident escalation path:
- Number of incidents in the last 12 months:

4. Change management

- Change triggers requiring review:
- Last major change date & review outcome:

9.7 Signals: metrics that show governance is working

To avoid governance becoming invisible or purely qualitative, you need **metrics**—not only on model performance but on governance itself. Useful governance metrics for AI modernization might include:

- Number of AI systems by risk tier.
- Percentage of systems with complete documentation.
- Time from submission to governance decision (by risk tier).
- Number and severity of AI-related incidents (over time).
- Percentage of high-risk systems with up-to-date reviews.
- Actions taken after governance reviews (e.g., conditions imposed, changes made).

These metrics tell you whether governance is practical, timely, and improving risk posture.

Manager checklist: governance operating model health

Use this checklist to assess your current state:

9.8 Governance Operating Model Health Checklist

- Structure
    - We have defined oversight bodies (e.g., Steering Committee, Review Board) with clear charters.
    - AI roles and responsibilities are described and communicated (business owners, model owners, system owners, risk/compliance).
- Flow
    - There is a documented AI governance workflow from intake to retirement.
    - Teams know which steps apply to their projects based on risk.
- Controls
    - We have a control matrix that links risk tiers to required controls.
    - Controls are embedded in project templates (not separate, hidden documents).
- Signals
    - We track governance metrics (e.g., review coverage, incident trends, documentation completeness).
    - Governance metrics are reviewed periodically by leadership.

If most of these are "no" or "not sure," your operating model is informal—and fragile.

Governance Model Health Gauge

9.9 Right-sizing governance: matching model to organization

There is no one-size-fits-all model. Over-engineering governance for a 150-person agency will suffocate progress; under-engineering for a multinational bank will invite chaos. Consider:

- **Size and complexity:** How many AI systems? How many business lines?
- **Regulatory exposure:** Highly regulated sectors require more formal governance.
- **Culture:** Are decisions centralized or distributed? How mature are risk processes today?

You can start light: a single AI Review Board, a simple intake process, and a few core metrics. As your AI footprint grows, you can decentralize—with domain-specific governance cells operating under a central standard.

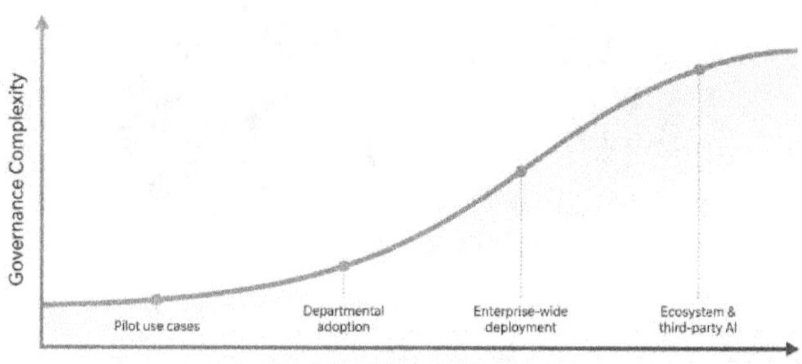

Governance Model Scaling Curve

Template: AI governance model design canvas

Use this as a design tool with your peers:

9.10 AI Governance Operating Model – Design Canvas

1. Context

- Organization size and structure:
- Main AI domains (e.g., customer, operations, risk, HR):
- Regulatory environment (short description):

2. Structure

- Central bodies (names and purposes):
- Domain or local governance units (if any):
- Key roles (business, model, system, risk, executive) and who fills them:

3. Flow

- Standard stages: [Intake, Assessment, Review, Deployment, Monitoring, Change, Retirement]
- Variations by risk tier:

4. Controls

- Control matrix summary (per risk tier):
- Minimum artifacts required:

5. Signals

- Governance KPIs (3–5 prioritized):
- Reporting cadence and audience:

Case pattern: central standards, federated execution

Many organizations succeed with a **"central standards, federated execution"** model:

- A central team defines AI policies, templates, risk frameworks, and governance structures.
- Business units or domains apply those standards locally, with their own review forums and domain expertise.

- High-impact or cross-cutting systems still go to a central AI Review Board or Steering Committee.

For IT managers, this model means:

- You do not try to centralize every decision.
- You provide standard tools so local teams can run governance without reinventing it.
- You keep a global view through shared metrics and periodic portfolio reviews.

Central Standards, Federated Execution

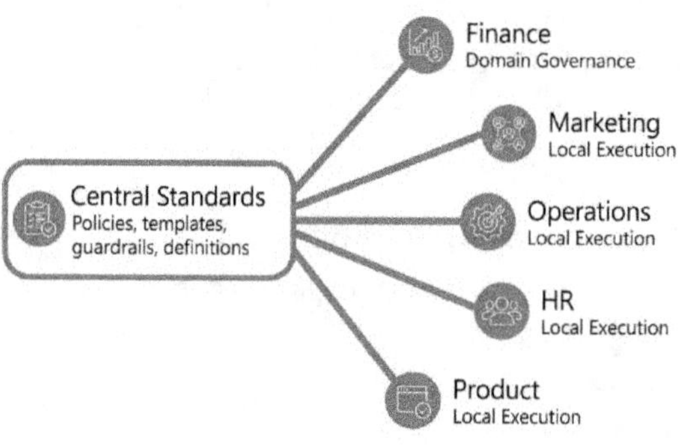

9.11 The Manager's Governance Playbook (90-day plan)

To move from ad-hoc to deliberate governance, apply this 90-day playbook:

0–30 days: Map what you already have

- List all AI-relevant committees, forums, and roles.
- Map current processes: how AI ideas come in, how they are approved, how issues are handled.
- Identify duplication, gaps, and places where decisions stall.

31–60 days: Define your operating model

- Draft an AI governance workflow for all AI projects.
- Define or refine the AI Review Board charter and membership.
- Create the AI governance control sheet template and risk-tiered control matrix.

61–90 days: Pilot and refine

- Select 2–3 active AI projects to follow the new governance workflow.
- Run the projects through intake, assessment, review, and post-deployment check.
- Collect feedback from teams and adjust templates, cadence, and roles as needed.

9.12 The AI Reality Check: governance that actually changes behavior

You know your governance operating model is working **only** if it changes behavior:

- High-risk AI projects look different (more controls, more oversight) than low-risk ones.
- Some projects are delayed, re-scoped, or stopped for governance reasons—and that is accepted.
- Incidents are handled faster and more consistently because roles and paths are clear.
- Teams prepare better, earlier, because they know what governance will expect.

If none of this is happening, you likely have "governance theater"—documentation without decisions, meetings without consequences.

Governance Theater vs Governance in Practice

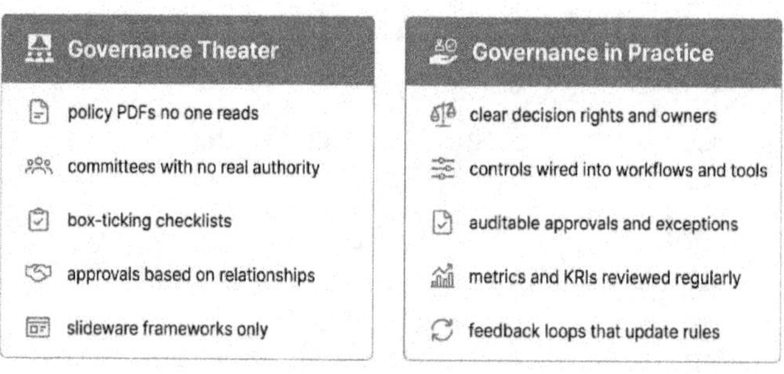

9.13 The Monday Morning Test

On Monday morning, a serious AI manager should be able to answer:

1. Can I sketch, from memory, our AI governance workflow from idea to retirement—and would my teams sketch roughly the same thing?
2. For my top AI systems, do I know which governance body approved them, when, and under what conditions?
3. If a new high-impact AI proposal arrives this week, do I know exactly which forms, assessments, and reviews it must go through?
4. Do we have metrics that show how many AI systems are governed, how many have up-to-date reviews, and how long approvals take?
5. When was the last time governance caused a project to change direction—and did that ultimately improve our risk posture or outcomes?

If some answers are "not yet," that is your design brief. An AI governance operating model is not about creating more meetings. It is about giving your modernization program **operational clarity** so you can scale AI with confidence, not hope.

Governance Operating Model Maturity Gauge

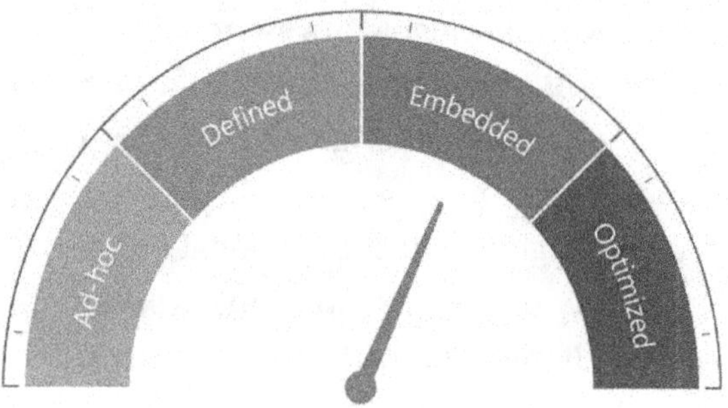

10 Embedding Responsibility by Design

Opening vignette

You are leading a major AI rollout: a mix of vendor models and in-house systems embedded into customer service, underwriting, and internal knowledge tools. You have set up an AI Review Board, written policies, and run a few bias audits. Still, you notice a pattern: governance conversations happen late—at pre-deployment review or, worse, after an incident. Teams see responsibility as something external they "pass through" rather than something they design for. In a portfolio review, your COO asks, "How do we know responsibility is baked into how we build AI, not just checked at the end?" That is the moment it clicks: you do not just need responsible AI governance— you need **responsibility by design** inside the day-to-day delivery machinery.

10.1 Why responsibility must live inside the delivery engine

If responsibility lives only in policies and final reviews, it loses the race against speed. Real-world AI projects move quickly: data scientists iterate on models, engineers wire up integrations, and product teams experiment with generative features. If responsible AI checks arrive only at "go-live," they show up too late, when the architecture and expectations are already set.

Responsibility by design means:

- Risks and ethics are considered **when the problem is framed**, not just when code is ready.
- Procurement, architecture, and sprint planning include **responsibility requirements** as first-class items.
- Continuous monitoring and feedback loops are designed **before deployment**, not bolted on afterward

Before vs After: Responsibility Placement

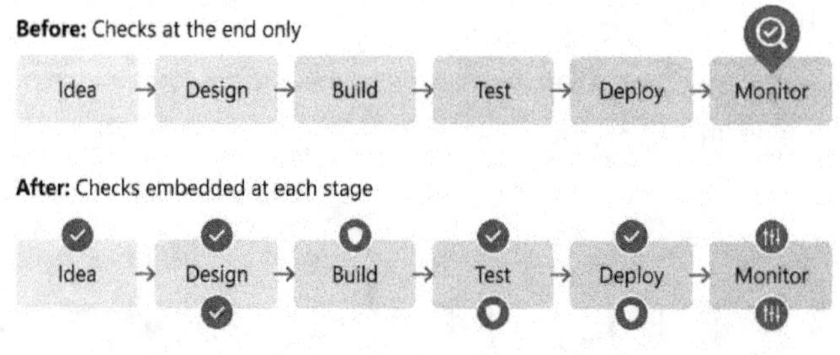

10.2 Responsibility by design: the core concept

Responsibility by design extends ideas such as "privacy by design" and "security by design" throughout the full AI lifecycle. The idea is simple:

- Every stage of the lifecycle has explicit responsibility checkpoints.
- Each checkpoint has clear questions, artifacts, and owners.
- Those checkpoints are integrated into existing processes (SDLC, procurement, change management), not added as parallel bureaucracy.

The outcome is an AI delivery system where responsible behavior is the default, not a special intervention.

Responsibility by Design Loop

10.3 The Responsibility by Design lifecycle

For IT managers, a practical lifecycle with embedded responsibility looks like this:

1. **Scope & intake:** Clarify purpose, stakeholders, risks, and suitability of AI.
2. **Design & procurement:** Set responsibility requirements for data, models, vendors, and workflows.
3. **Build & validate:** Implement controls and test against responsible AI criteria (fairness, safety, transparency, compliance).
4. **Deploy & enable:** Configure safe defaults, disclosures, oversight, and training.
5. **Monitor & adapt:** Continuously monitor behavior, drift, incidents, and user feedback—and feed changes back into design.

Each stage gets its own checklist and, for higher-risk systems, formal checkpoints.

Stage Responsibility by Design Lifecycle

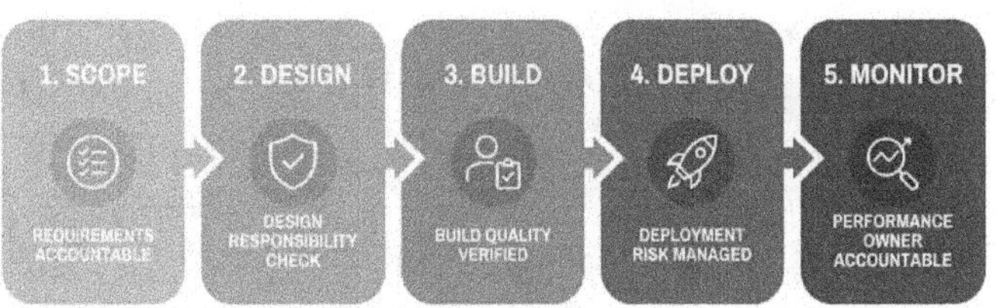

Stage 1: Scope & intake – responsible problem framing

Responsibility starts when you decide **what problems** to apply AI to and **how** you frame them. Poor scoping leads directly to irresponsible outcomes, even if the model is technically strong.

Questions to embed in intake:

- Is AI the right tool for this problem, or is there a simpler, less risk-heavy solution?
- Who benefits if this system works well—and who could be harmed if it does not?
- Does the use case affect rights, access to services, employment, safety, or vulnerable groups?
- What existing inequities or sensitivities surround this process today?

These become mandatory fields in your AI Project Intake & Policy Trigger form (from earlier chapters), so they are answered **before** resources are committed.

10.4 Scope & intake responsibility checklist

- Purpose aligns with mission and values.
- AI is justified over non-AI alternatives.
- Stakeholders and vulnerable groups identified.
- Potential harms clearly listed (safety, fairness, misuse, privacy).
- Preliminary risk tier assigned and reviewed.

Stage 2: Design & procurement – baking responsibility into requirements

Once you decide to proceed, design and procurement are where responsibility becomes concrete. This is where you turn principles into **requirements**:

For internal builds:

- Data requirements include representativeness, consent/legality, and data minimization.
- Architecture requirements include logging, explainability hooks, and human-in-the-loop points.
- Risk requirements include safety/fairness tests and monitoring metrics defined up front.

For vendor procurement:

- RFPs/contract clauses specify expectations for fairness, safety, explainability, documentation, monitoring, and incident support.
- Vendors must provide artifacts (e.g., model cards, risk documentation, monitoring capabilities) as part of selection and onboarding.

10.5 Design & procurement responsibility checklist

- Responsibility requirements included in design documents and user stories.
- Data governance requirements defined (sources, quality, access, retention).
- Model and system must expose necessary logging and explainability data.
- Vendor requirements for responsible AI are documented (if applicable).
- Planned checkpoints and artifacts agreed with stakeholders (risk, legal, business).

Design & Procurement Responsibility Bridge

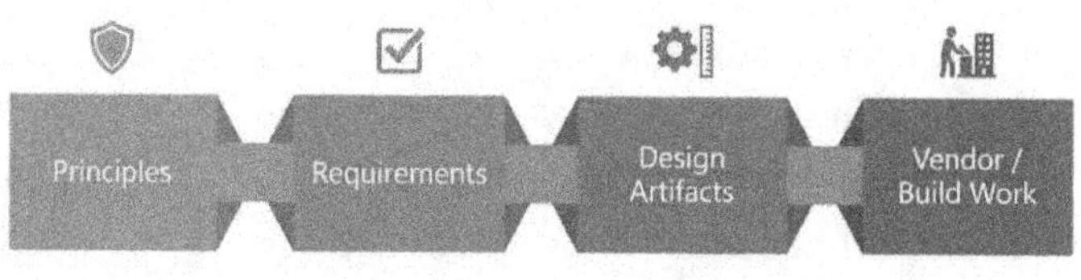

Template: Responsible AI design brief (per project)

Embed this into your standard design process:

10.6 Responsible AI Design Brief – Template

1. Purpose & context

- Problem we are solving:
- Why AI is appropriate:
- Stakeholders and affected groups:

2. Responsibility requirements

- Fairness:
  - Target fairness goals and metrics:
  - Groups or segments to monitor:
- Safety & misuse:
  - Prohibited behaviors and misuse scenarios to guard against:

- o Human-in-the-loop requirements (when humans must review/override):
- Transparency & explainability:
  - o Required explanation level for different audiences (users, staff, regulators):
  - o Logging and documentation expectations:
- Privacy & security:
  - o Data minimization rules:
  - o Sensitive attributes handling:

3. Architecture & workflow implications

- Where the AI will sit in the workflow:
- What decisions it makes vs recommends:
- Integration points for logging, monitoring, and escalation:

4. Governance & checkpoints

- Planned review forums and dates:
- Artifacts to produce (model card, fact sheet, risk assessment, test results):

(Insert Diagram 6: "Responsible Design Brief Layout" – a blue page sketch illustrating the main sections.)

Stage 3: Build & validate – responsibility in development and testing

In development, responsibility-by-design manifests as **specific checks in your SDLC and MLOps pipelines**. Rather than only testing technical performance, you test:

- Fairness: performance on key groups, disparity metrics, and bias tests.
- Safety: harmful content, risky failure modes, and misuse resilience.
- Robustness: performance under noise, distribution shifts, and adversarial prompts.
- Compliance: alignment with policy and regulatory constraints.

For IT managers, this means adding responsibility gates to CI/CD, model approval workflows, and test plans—not running separate, ad hoc reviews.

Build & validate responsibility checklist

- Test plan includes fairness, safety, robustness, and misuse scenarios.
- Automated checks (where possible) are wired into CI/CD or MLOps pipelines.
- Model card and AI fact sheet are updated with test results and limitations.
- Risk assessment is updated with new insights and residual risks.
- Go/no-go criteria include responsibility-related thresholds (not just accuracy).

Responsibility-Aware SDLC

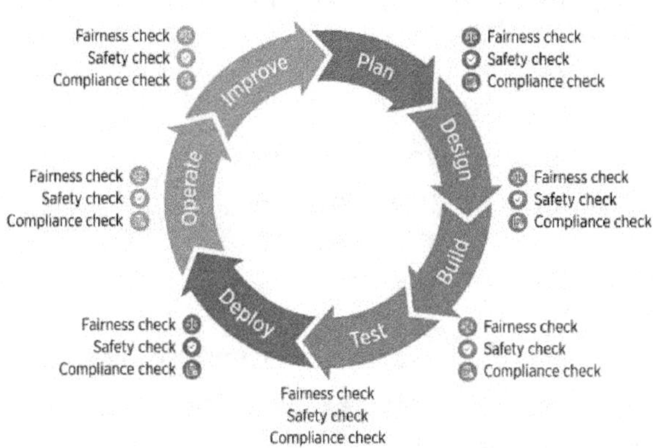

Template: Responsible AI test plan addon

Attach this to existing test plans:

Responsible AI Test Plan – Add-On Template

1. Fairness tests

- Groups/segments evaluated:
- Metrics (e.g., error rates, selection rates across groups):
- Acceptance thresholds and rationale:

2. Safety & misuse tests

- Harm scenarios tested (e.g., harmful outputs, dangerous recommendations):

- Misuse scenarios tested (e.g., prompt injection, data exfiltration, abuse of generative capabilities):
- Observed behavior and mitigation actions:

3. Robustness tests

- Stress tests and edge cases tried:
- Behavior under degraded or shifted data:

4. Compliance tests

- Checks against policy requirements (e.g., no prohibited attributes, documentation completeness):
- Privacy checks (data minimization, masking, consent assumptions):

5. Results & decision

- Summary of findings:
- Issues requiring mitigation before deployment:
- Residual risks accepted (by whom):

Stage 4: Deploy & enable – responsible rollout and human enablement

Deployment is where responsible design meets reality. Responsibility by design at this stage means:

- **Safe defaults:** Conservative thresholds, limited autonomation at first, clear escalation paths.

- **User disclosures:** Plain-language notices when AI is involved, with options for human review.
- **Training:** Frontline staff understand what the system does, its limits, and how to override or escalate.
- **Operational readiness:** Incident runbooks, escalation contacts, and monitoring dashboards in place on day one.

Deploy & enable the responsibility checklist

- System goes live with conservative thresholds and clear human-review rules.
- User interfaces include appropriate disclosures and recourse options.
- Staff training completed on capabilities, limits, and escalation paths.
- Incident runbook and escalation plan confirmed and tested.
- Initial post-deployment review scheduled (e.g., 30–90 days).

Responsible Deployment Readiness

Must-Have Before Go-Live

- ☑ Governance approvals completed
- ☑ Risk assessment and DPIA (if applicable)
- ☑ Fairness and bias testing passed
- ☑ Safety and security review complete
- ☑ Compliance and legal sign-off
- ☑ Human-in-the-loop and escalation paths defined

Post-Launch Commitments

- ☑ Continuous monitoring in place
- ☑ Drift and performance alerts configured
- ☑ Incident and rollback playbook ready
- ☑ Periodic re-assessment scheduled
- ☑ Feedback channels for users
- ☑ Review of logs and audit trails

Stage 5: Monitor & adapt – continuous responsibility in production

Responsibility does not end at launch; that is when it is truly tested. Continuous monitoring should track:

- Technical performance (accuracy, latency, error rates).
- Fairness and safety metrics in production.
- Misuse attempts, security events, and anomalies.
- User feedback, complaints, and incident patterns.

Monitoring should be risk-tiered: high-impact systems receive more intense and frequent reviews; low-risk tools receive lighter oversight.

Monitor & adapt responsibility checklist

- Monitoring scope defined per risk tier (metrics, thresholds, frequency).
- Dashboards and alerts implemented and tested.
- Regular review cadences set (e.g., monthly for high-risk, quarterly for medium).
- Incidents and complaints logged, reviewed, and fed back into design.
- Retraining and change processes include updated risk and responsibility assessments.

Responsibility Monitoring Loop

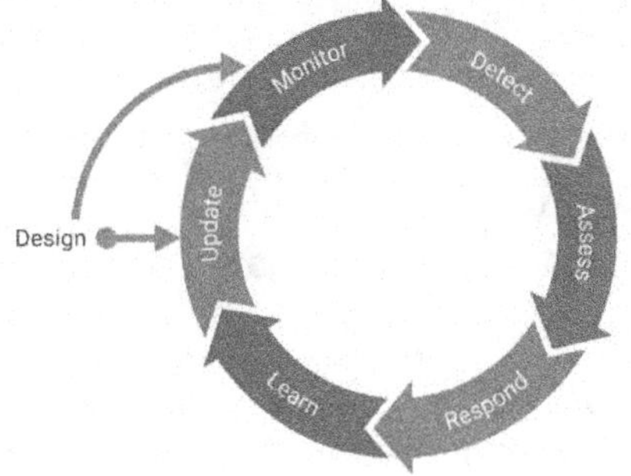

Responsible procurement: buying AI with responsibility built-in

For many modernization programs, a large portion of AI capability comes from vendors. Responsibility by design in procurement means:

- Defining responsible AI expectations as non-negotiable requirements.
- Asking vendors to supply their own model cards, risk assessments, and monitoring capabilities.
- Ensuring contracts include obligations for transparency, incident reporting, and updates.

Responsible AI procurement checklist

- RFPs and contracts include clear responsible AI criteria (fairness, safety, transparency, monitoring).
- Vendors provide documentation (model cards, testing summaries, risk information).
- Integration plan includes logging, monitoring, and explainability hooks.
- Vendors change and update processes are documented and reviewed.
- Exit strategy defined if the vendor cannot meet the responsible AI expectations.

"Responsible Procurement Flow"

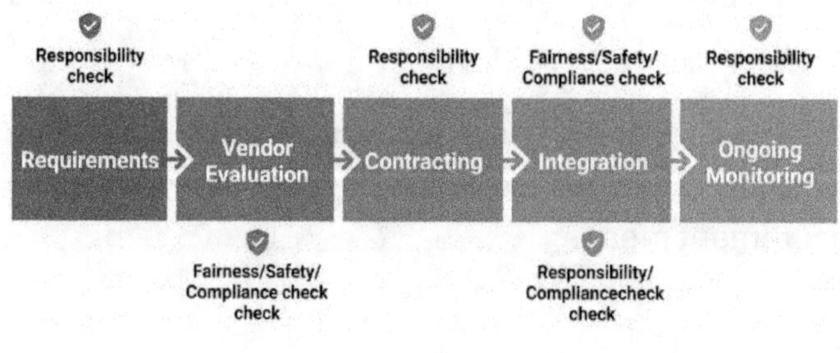

Template: Responsible AI vendor questionnaire (short form)

Use this with vendors as part of selection or onboarding:

10.7 Responsible AI Vendor Questionnaire – Template

1. System overview

- Description of AI functionality:
- Intended use cases and limitations:

2. Data & model

- Data sources and governance practices:
- Training and evaluation approach (high level):

3. Fairness & safety

- Fairness tests performed and key results:
- Safety/misuse scenarios tested and mitigations:

4. Transparency & documentation

- Available documentation (model card, fact sheet, technical docs):
- Explainability features provided to customers:

5. Monitoring & incidents

- Monitoring capabilities (metrics, alerts, dashboards):

- Incident response process and support commitments:

6. Compliance & certifications

- Relevant regulatory or framework alignments:
- Any third-party assessments or certifications:

10.8 Integrating responsibility into existing governance processes

You already have governance processes: architecture review boards, change advisory boards (CABs), security reviews, and procurement workflows. Responsibility by design means **piggybacking** on these rather than inventing entirely new machinery.

Practical moves:

- Add responsible AI fields into existing architecture review templates.
- Include responsible AI checkpoints on CAB forms (e.g., "Is this change AI-related? Does it alter risk tier or controls?").
- Ensure your AI Review Board uses the same risk ratings and documentation patterns as your enterprise risk function.

This keeps responsibility "in the flow" rather than creating parallel tracks that teams will try to avoid.

Template: Responsible AI add-on to project governance

Add this page to your existing project governance templates:

Responsible AI Governance Add-On – Template

1. AI involvement

- Does this project introduce or significantly change AI functionality? [Yes/No]
- AI systems affected (names):

2. Responsibility scope

- Risk tier(s) of AI systems involved:
- Responsibility focus areas (tick all that apply): [Fairness] [Safety] [Transparency] [Privacy] [Misuse] [Compliance]

3. Required artifacts

- Model card: [Required / N/A] – Status:
- AI fact sheet: [Required / N/A] – Status:
- Risk assessment: [Required / N/A] – Status:
- Responsible for test plan results: [Attached / Pending]

4. Approvals & conditions

- Governance forum(s) consulted:

- Key decisions and conditions:

5. Monitoring alignment

- Monitoring updates required: [Yes/No] – Description:

The Manager's Responsibility-by-Design Playbook (90-day plan)

To embed responsibility into your AI delivery engine, use this 90-day playbook:

0–30 days: Diagnose and align

- Map where responsible AI is considered today (intake, design, build, deploy, monitor).
- Identify gaps where responsibility is absent or only shows up late.
- Align with risk, legal, and architecture leads on the Responsibility by Design lifecycle.

31–60 days: Design the embeds

- Create the Responsible AI Design Brief, Test Plan Add-On, Vendor Questionnaire, and Governance Add-On templates.
- Integrate responsibility fields into existing intake, architecture review, CAB, and procurement forms.

- Define responsibility checkpoints for each lifecycle stage (for medium and high-risk systems).

61–90 days: Pilot and iterate

- Select 2–3 AI projects (mix of internal and vendor) to pilot the new embeds.
- Run them through the full lifecycle with responsibility checkpoints.
- Collect feedback, adjust templates, and refine thresholds to keep governance "light but real."

Responsibility-by-Design Roadmap

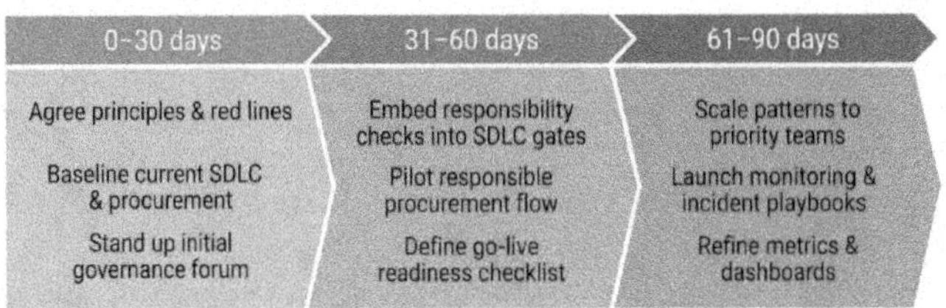

Figure 1 Example Tasks

10.9 The AI Reality Check: avoiding "responsibility theater."

Responsibility theater happens when organizations talk about responsible AI, publish principles, and run occasional reviews—but day-to-day workflows remain unchanged. Signs you are in responsibility theater:

- Responsibility is only discussed at final approvals, not at intake or design.
- Templates exist but are rarely used or filled out meaningfully.
- Vendor AI is onboarded without responsible AI questions.
- Incidents lead to blame but not to updates in design and processes.

Responsibility by design is the antidote: if governance is not changing tickets, designs, tests, and deployments, it is not yet real.

The Monday Morning Test

On Monday morning, a serious AI manager should be able to answer:

1. At which points in our AI lifecycle do we have explicit responsibility checkpoints—and who owns them?
2. For our current high-impact AI projects, can I point to a Responsible AI Design Brief and Test Plan that shaped real decisions, not just documentation?

3. When we buy AI from vendors, do we consistently ask responsibility-related questions and act on the answers?
4. Do our deployment processes enforce safe defaults, disclosures, and training before going-live for AI systems?
5. How do monitoring insights and incidents feed back into design and procurement decisions, not just into firefighting?

If your answers are mixed or vague, that is your design agenda. Responsibility by design turns responsible AI from a compliance topic into **a way of building**, giving your modernization program both speed and integrity.

11 Data Integrity and Stewardship

Opening vignette

You have several AI systems live: a claims triage model, a customer-churn predictor, and a generative assistant for internal knowledge. Performance looks strong—until a pattern emerges. The fraud model's false-positive rate spikes following a change to the data pipeline. A regulator questions the data source in a credit model. A senior leader asks a simple question: "Where did this data come from, who approved its use, and how do we know it is still accurate?" Your team scrambles across dashboards, SQL queries, and email threads. Nobody can show a clean, end-to-end   story.
In that moment, you realize the bottleneck is not modeling capability; it is **data integrity and stewardship**. Without trustworthy data and clear ownership, Responsible AI is built on sand.

11.1 Why data integrity is the foundation of Responsible AI

AI systems are only as trustworthy as the data feeding them. If data is incomplete, biased, stale, or poorly controlled, even the most advanced model will produce unreliable or unfair outcomes. In the AI era, data integrity must cover:

- **Fairness:** Are different groups represented appropriately, and are we aware of historical biases?
- **Privacy:** Was the data collected and used with proper consent and a lawful basis?
- **Performance:** Is the data accurate, timely, and fit for the model?

For IT managers, this shifts data from "back-end plumbing" to a **strategic asset and risk vector** that must be governed as deliberately as code or infrastructure.

Data as the Intersection

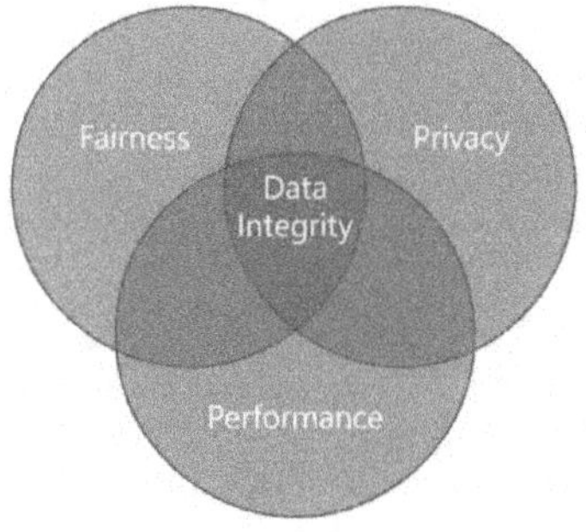

11.2 What "data governance" means in the AI era

Traditional data governance focuses on catalogs, access policies, and compliance reporting. In the AI era, data governance became **operational and model-centric**: it must explain how data flows from source systems through pipelines into models, and how quality and permissions are managed along the way.

Core elements include:

- **Policies and standards:** Rules for collection, use, retention, and quality of data used in AI.
- **People and roles:** Data owners, stewards, custodians, and producers with clear responsibilities and RACI.
- **Processes:** Workflows for onboarding new datasets, approving data use, handling quality issues, and responding to incidents.
- **Technology:** Lineage tracking, catalogs, quality monitoring, and access control tooling.

Four Pillars of AI-Era Data Governance

Policies	People
• Usage & access rules • Retention & deletion standards	• Data owners & stewards • Training & accountability
Processes	**Technology**
• Data quality & lineage workflows • Issue & change management	• Catalogs & governance tools • Security & monitoring platforms

11.3 Data lineage: knowing where your AI's data comes from

Data lineage answers two critical questions:

- **Upstream:** Where did this data originate, and what transformations has it gone through?
- **Downstream:** Which models and reports depend on it?

For AI systems, robust lineage enables you to:

- Trace problems (bias, errors, drift) back to specific sources or transformations.
- Understand the impact when you change or retire a data source.
- Demonstrate to auditors and regulators how data was prepared for high-risk models.

AI-ready lineage is granular, continuously updated, and accessible to both humans and AI systems.

Data lineage checklist (per key dataset/model)

- Source systems documented (including owners and system of record).
- Transformations and joins are documented at a level that engineers and reviewers can understand.
- Downstream models and reports identified.
- Lineage is kept up to date as pipelines change.

End-to-End Data Lineage for an AI Model

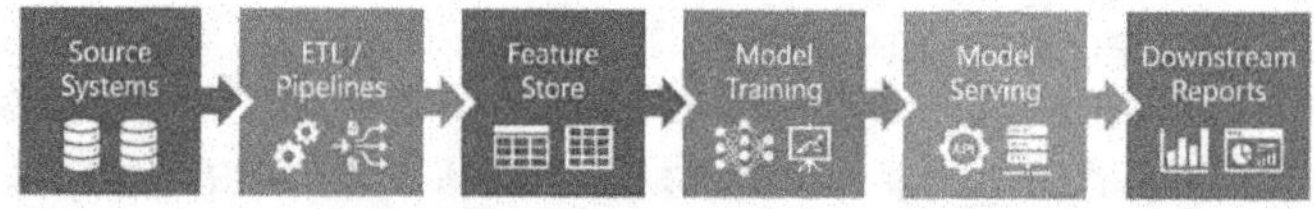

Consent, purpose, and lawful use

AI modernization often repurposes data collected for other reasons. In regulated environments and privacy-aware cultures, you must ensure:

- Data was collected with appropriate consent or lawful basis for its current use.
- The **purpose** of use is compatible with the original collection purposes.

- Special categories of data (health, biometrics, minors, etc.) are handled with heightened safeguards.

This is not just a legal issue; misusing data undermines trust and can lead to costly rollbacks.

Consent & purpose checklist

- For each dataset, the original collection purpose and consent status are documented.
- AI use cases involving the data are reviewed for purpose compatibility.
- Sensitive data categories are flagged with additional controls and approvals.
- Data minimization is applied: only necessary fields are used.

Data Use Decision Tree

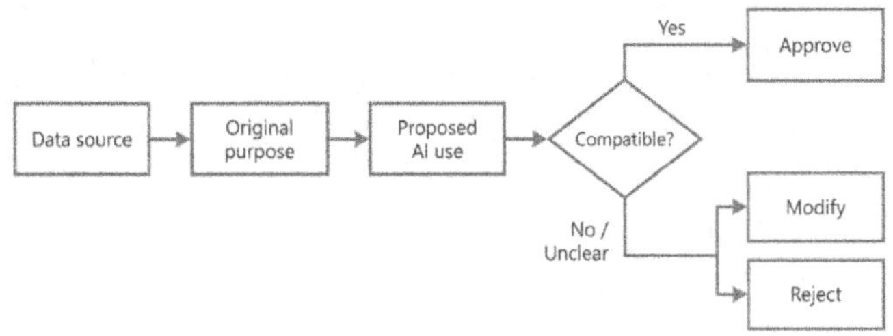

11.4 Labeling quality and annotation integrity

Supervised models and many evaluation processes depend on labeled data. Poor labeling quality leads to noisy models, hidden bias, and brittle behavior. Labeling risk increases when:

- Labelers are poorly trained or misaligned on definitions.
- Labeling guidelines are ambiguous or incomplete.
- There is no quality assurance sampling or inter-rater reliability checks.

For IT managers, you must treat labeling as a **governed process**, not a one-time task.

Labeling quality checklist

- Clear labeling guidelines exist with examples and edge cases.
- Labelers receive training and can ask questions.
- Quality checks (spot checks, agreement metrics) are in place.
- Feedback loops exist from model performance back to labeling practices.

Labeling Quality Loop

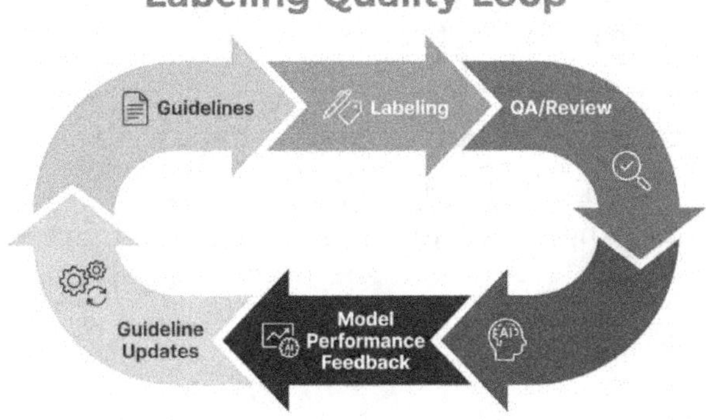

Access control and "least privilege" for AI data

AI initiatives often create pressure to "open up" data. Without discipline, this leads to over-exposure of sensitive information. Sound access control for AI data should include:

- **Role-based access control**: roles and datasets mapped; no generic "super user" access except tightly controlled.
- **Least privilege**: teams get only the data they reasonably need (often through curated views or feature stores).
- **Segmentation**: separation between training, test, and production data; special handling for PII/PHI.
- **Logging**: access is logged and can be audited for anomalies.

Access control checklist (AI datasets)

- Data access defined per role and approved by data owner or steward.
- Sensitive attributes and fields are identified and appropriately masked or restricted.
- Training and test data are segregated and protected.
- Access logs are retained and periodically reviewed.

AI Data Access Layers

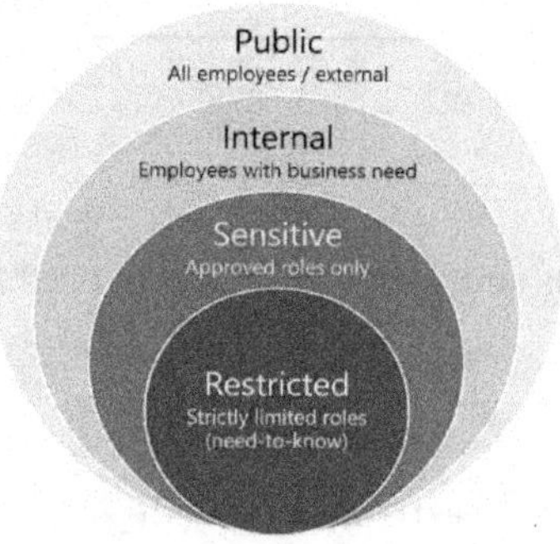

11.5 Data trustworthiness: defining quality for AI

You cannot ensure data integrity without defining **"good enough"** for your AI use cases. Common data quality dimensions include:

- **Accuracy:** Data reflects real-world values correctly.
- **Completeness:** Required fields and records are present.
- **Timeliness:** Data is recent enough for decisions.
- **Consistency:** No conflicting values across systems.
- **Validity:** Data conforms to expected formats and ranges.

For AI, you add:

- **Representativeness:** Data adequately covers the populations and scenarios the model will see.
- **Stability:** Data distributions do not change unexpectedly, or such changes are detected.

Data quality checklist (AI-critical datasets)

- Quality dimensions and thresholds defined for each critical dataset.
- Automated checks run regularly (nulls, ranges, referential integrity, distribution shifts).
- Quality issues are logged, triaged, and assigned owners.
- AI model owners are notified when significant quality issues arise.

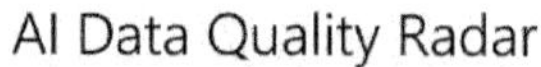
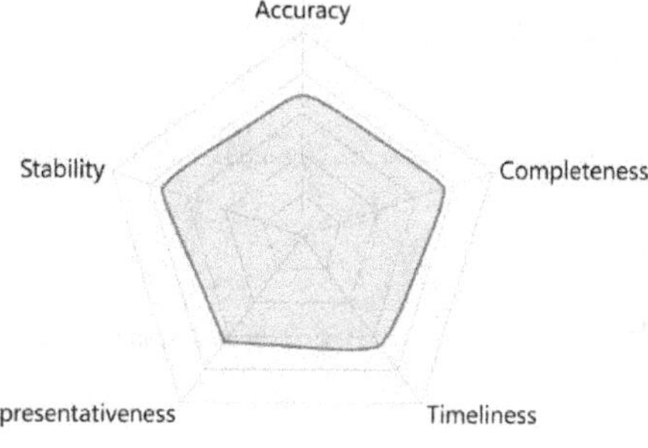

11.6 Introducing the Data Stewardship Board

To make data governance real, you need a **cross-functional oversight body** focused on data: the **Data Stewardship Board**. This is not about bureaucracy; it is about having the right people making decisions about data that powers AI.

Typical composition:

- Data owners from major domains (finance, operations, customer, HR, etc.).
- Data stewards who manage data quality and standards day-to-day.
- Representatives from IT, security, legal/privacy, compliance, and AI/ML.
- Optional: business and program leaders for high-impact domains.

Core responsibilities:

- Approve data standards, quality thresholds, and critical dataset definitions.
- Prioritize data quality and governance remediation initiatives.
- Resolve data ownership conflicts.
- Review data-related risks and incidents tied to AI systems.

Data Stewardship Board in the Governance Stack

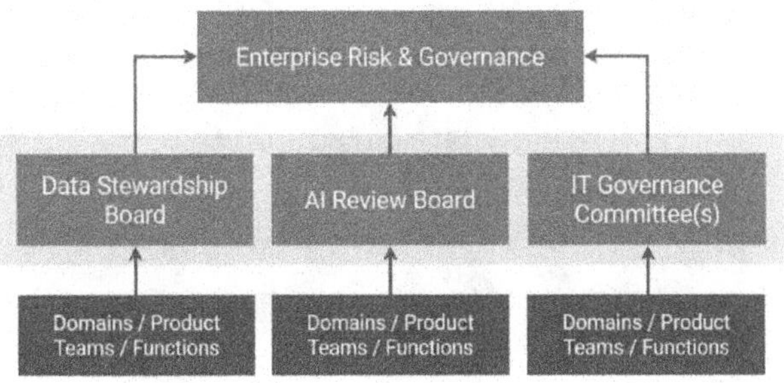

Template: Data Stewardship Board charter

Data Stewardship Board Charter – Template

1. Purpose

- Ensure that data used in AI and analytics is trustworthy, well-governed, and aligned with policies and regulations.

- Provide cross-functional oversight on data quality, access, and use.

2. Scope

- Critical datasets used in high-impact AI systems.
- Data standards, definitions, lineage practices, and quality management.
- Data-related incidents and risks affecting AI systems.

3. Membership

- Chair: [Role/title]
- Core members:
    - Data owners (key domains):
    - Data stewards:
    - IT/data platform representative:
    - Security/privacy representative:
    - AI/ML representative:
    - Compliance/risk representative:

4. Responsibilities

- Approve and maintain data standards and quality thresholds.
- Approve critical dataset catalog and ownership.
- Review lineage coverage and gaps for AI-critical data.
- Review significant data quality or misuse incidents and ensure remediation.

5. Cadence & operations

- Meeting frequency (e.g., monthly).
- Required input (quality reports, incident summaries, AI system changes).
- Decision recording and escalation paths.

11.7 Defining data roles: owners, stewards, custodians

Clarity on data roles makes everything else easier. Typical roles:

- **Data owner:** Senior leader accountable for a data domain's quality, use, and compliance. Approves access policies and major changes.
- **Data steward:** Operational role managing day-to-day quality, metadata, and enforcement of standards for specific datasets.
- **Data custodian:** IT/platform role responsible for the technical storage, security, and backup of data.
- **Data producer:** Teams or systems that generate data.
- **Data consumer:** Teams (including AI and analytics) that use data.

Data roles checklist (per key domain/dataset)

- Data owner identified (decision-making and accountability).
- Data steward identified (day-to-day management).
- Data custodian identified (technical operations).
- RACI documented for key decisions (access, quality, structural changes).

Practical methods for maintaining data trustworthiness

To keep data trustworthy in practice, you need **continuous operational mechanisms**:

- **Critical dataset catalog:** Maintain a list of datasets considered "critical" for AI systems (with owners, stewards, and quality expectations).
- **Issue management:** Central process for logging, triaging, and resolving data issues, linked to AI systems affected.
- **Quality SLAs:** Service-level agreements for data quality dimensions (e.g., maximum null rate, timeliness) for critical data.
- **Lineage coverage:** Targets (e.g., 100% lineage for critical datasets; partial for others).
- **Automated checks:** Data quality and lineage monitoring integrated into data pipelines.

11.8 Data trustworthiness checklist (portfolio level)

- Critical dataset catalog exists and is maintained.
- Data quality SLAs are defined for critical datasets.
- Data issues are tracked with clear owners and deadlines.
- Lineage coverage targets are set and monitored.
- AI model owners can see data quality and lineage information for their systems.

Data Trustworthiness Operating Loop

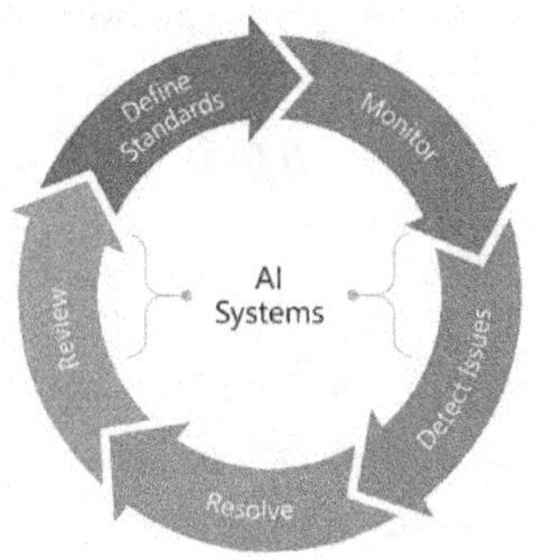

Template: Critical dataset record

Use this for each AI-relevant critical dataset:

Critical Dataset Record – Template

- Dataset name:
- Business domain:
- Description and main use cases (including AI models):

1. Ownership & stewardship

- Data owner (name, role):
- Data steward (name, role):
- Data custodian (team):

2. Quality & SLAs

- Key quality dimensions and targets (e.g., accuracy, completeness, timeliness):
- Representative considerations (key segments/groups):
- Monitoring frequency and methods:

3. Lineage & access

- Source systems and major transformations:
- Downstream AI models and reports:
- Access policy summary (roles, restrictions, masking rules):

4. Issues & incidents

- Known quality issues and remediation plans:
- Data-related incidents affecting AI systems (past 12–24 months):

Manager checklist: data integrity readiness for AI systems

At the system level, you can quickly assess data integrity readiness:

Data Integrity Readiness Checklist (per AI system)

- Data sources documented with owners and stewards.
- Lineage from sources to model inputs is known and accessible.
- Training and evaluation datasets meet minimum quality thresholds.
- Consent and lawful-use checks completed for key data.
- Access to training and production data is role-based and logged.
- Data quality monitoring is in place and connected to model monitoring.
- Data-related incidents can be traced and escalated via clear paths.

The Manager's Data Stewardship Playbook (90-day plan)

To embed data integrity and stewardship into your modernization program, use this 90-day playbook:

0–30 days: Map and prioritize

- Inventory AI-critical datasets (those feeding high-impact models).
- Identify current data owners, stewards, and custodians—or gaps.

- Assess current lineage coverage and quality monitoring for these datasets.

31–60 days: Establish structures and standards

- Define or refine data roles and responsibilities (owners, stewards, custodians).
- Draft the Data Stewardship Board charter and confirm membership.
- Create the Critical Dataset Record template and start populating it for top datasets.

61–90 days: Operationalize

- Set quality SLAs and monitoring plans for critical datasets.
- Integrate data quality and lineage status into AI system reviews.
- Run a joint session between the Data Stewardship Board and AI Review Board to review one or two systems end-to-end.

The AI Reality Check: avoiding "data governance theater."

Data governance theater happens when organizations create policies, name stewards, and buy catalogs—but nothing meaningful changes in how data is produced, managed, or used. Signs you are in data governance theater:

- Data roles are named on paper, but nobody knows what they do.
- Lineage exists in slides, not in the tools teams consult.
- Quality issues are discussed repeatedly, but lack owners and deadlines.
- AI incidents lead to "model fixes" without examining data root causes.

Real data integrity and stewardship are visible when:

- Data issues are logged and resolved with clear accountability.
- AI model owners routinely consult lineage and quality views.
- Data standards and SLAs drive pipeline design and remediation.

The Monday Morning Test

On Monday morning, a serious AI manager should be able to answer:

1. For my highest-impact AI systems, can I identify the critical datasets they rely on and the people who own and steward them?
2. Can I see, in one place, where that data comes from (lineage) and which systems it feeds, including AI models?

3. Do we have defined quality expectations and monitoring for those datasets—and do model owners see those signals?
4. How do we ensure the data we use for AI respects consent, purpose, and lawful use, especially for sensitive categories?
5. When a data-related incident occurs, is there a clear path through the Data Stewardship Board or similar structure to resolve root causes?

If your answers are partial or uncertain, that is your agenda. Strong data integrity and stewardship turn your data from a hidden liability into a **trusted foundation** for Responsible AI—one that supports fairness, privacy, performance, and governance at scale.

12 Culture, Training, and Human Oversight

Opening vignette

You have invested heavily in AI governance: policies are published, an AI Review Board is in place, and risk templates are in place. However, incidents keep catching you off-guard. A frontline team quietly turns off an AI recommendation because it "gets in the way." An engineer ships an unapproved prompt change to "make the chatbot sound friendlier." A product owner dismisses a fairness concern as "edge cases we do not have time for." In an internal post-mortem, a senior leader says, "On paper, our governance looks great. So why are we still surprised?" That is when it becomes clear: governance does not live in documents; it lives in people. Without the right culture, training, and human oversight, Responsible AI is a PowerPoint fantasy.

12.1 Why culture matters as much as controls

You can design excellent processes, frameworks, and templates. If the culture quietly rewards speed over safety, silence over escalation, and clever hacks over documented change, your governance will be performative. Culture is **what people do when no one is watching**.

For IT managers leading AI modernization, a responsibility culture shows up when:

- Teams raise concerns early, even when it slows delivery.
- Engineers and product owners see ethics, fairness, and safety as part of their job—not "someone else's problem."
- Leaders react to red flags with curiosity and support, not blame.

Your job is to shape conditions where responsible behavior is the path of least resistance, not an act of heroism.

Four Elements of Responsibility Culture

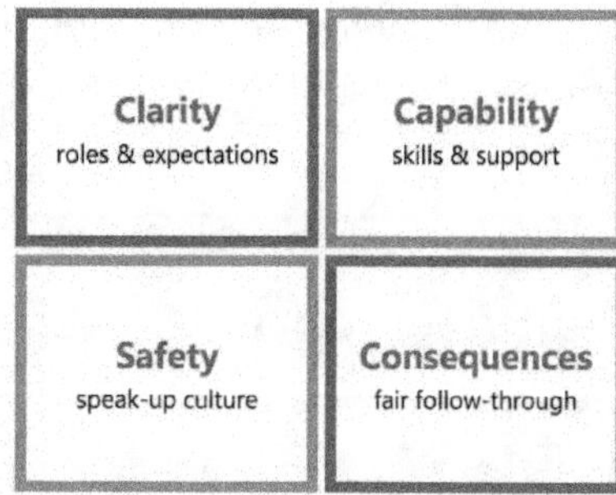

12.2 Elements of a "responsibility culture."

A responsibility culture has four core elements:

1. **Clarity:** People know the principles, policies, and expectations around AI.
2. **Capability:** People have the skills and tools to act responsibly (checklists, playbooks, contacts).
3. **Psychological safety:** People feel safe speaking up about risks or uncertainty.
4. **Consequences:** Responsible behavior is recognized; irresponsible shortcuts face consequences.

As a manager, you cannot force values, but you can consistently reinforce these elements through decisions, communication, and incentives.

Building ethical literacy across teams

Ethical literacy is not about turning everyone into philosophers. It is about giving people enough understanding to recognize when an AI decision might be problematic, and what to do next.

Focus on teaching:

- Common AI failure modes (bias, hallucinations, misuse, over-reliance on automation).

- Real examples from your own domain (e.g., unfair scoring, harmful advice, opaque denials).
- How your organization defines fairness, safety, and acceptable tradeoffs.
- The escalation paths and forums where ethical concerns are handled.

Short, scenario-based sessions are more effective than long lectures. People remember stories and checklists, not abstract definitions.

Ethical Literacy Building Blocks

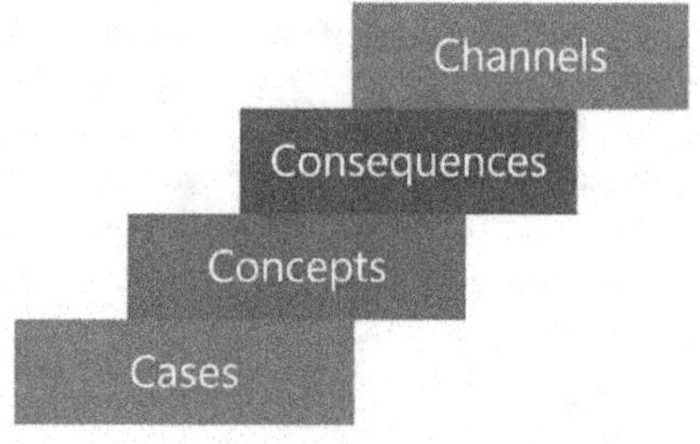

12.3 Training: making governance "how we work."

Training should be practical, role-specific, and recurring. You want to move from one-time awareness sessions to **embedded skill-building**.

Key audiences and training focuses:

- **Engineers and data scientists:** Responsible AI practices in data selection, modeling, testing, and documentation.
- **Product owners and business leads:** Use case selection, risk framing, recourse for users, communications.
- **Frontline staff:** How to interpret AI outputs, when to trust vs question, and how to escalate.
- **Managers:** How to lead conversations about trade-offs, handle escalations, and support responsible behavior.

12.4 Responsible AI training checklist

- Training modules exist for key roles (engineering, product, frontline, management).
- Sessions include concrete scenarios from your organization.
- Attendees know how to raise concerns and where to find tools/templates.
- Training is refreshed regularly, not just once at rollout.

Decision checklists: making the right thing easy

In high-pressure environments, people default to habits. Decision checklists provide teams with structured prompts at critical moments, design choices, go-live decisions, or incident triage.

Examples:

- A **design checklist** asking about stakeholder impacts, data quality, fairness, and potential misuse.
- A **deployment checklist** covering safe defaults, user disclosures, oversight, and training.
- An **incident checklist** about whom to notify, how to contain, and what to document.

Checklists are not bureaucracy; they are cognitive aids that support better judgment under constraints.

12.5 Decision checklist essentials

- Short and targeted to a specific decision or stage.
- Written in plain language.
- Integrated into existing tools (ticket templates, forms, runbooks).
- Owned and updated by named roles, not orphaned.

Template: Responsible AI decision checklist (design/review)

Use this in design reviews and risk assessments:

Responsible AI Design & Review Checklist – Template

For each AI use case/system:

1. Purpose & impact
   - Is the purpose aligned with mission and values?
   - Who benefits, and who could be harmed?
2. Data & fairness
   - Are key groups represented adequately in the data?
   - Could the system amplify existing inequities?
3. Safety & misuse
   - What could go wrong in realistic scenarios?
   - How might someone intentionally misuse this system?
4. Transparency & recourse
   - Can we explain decisions at the level appropriate for users and regulators?
   - How can affected people challenge or appeal decisions?
5. Oversight & accountability
   - Who owns outcomes for this system?
   - What oversight mechanisms are in place (reviews, monitoring, kill switch)?

12.6 Human oversight: what it is and what it is not

"Human in the loop" has become a mantra, but it often hides weak oversight. Real human oversight requires:

- **Clear points of intervention:** Moments where humans can review, approve, override, or stop AI-driven actions.
- **Adequate information:** Humans see enough context and explanation to make an informed judgment.
- **Time and authority:** Humans have the time and organizational power to act on what they see.

Weak oversight is when humans are nominally in the loop but rubber-stamp outputs due to workload, lack of understanding, or cultural pressure to trust the machine.

Strong vs Weak Human Oversight

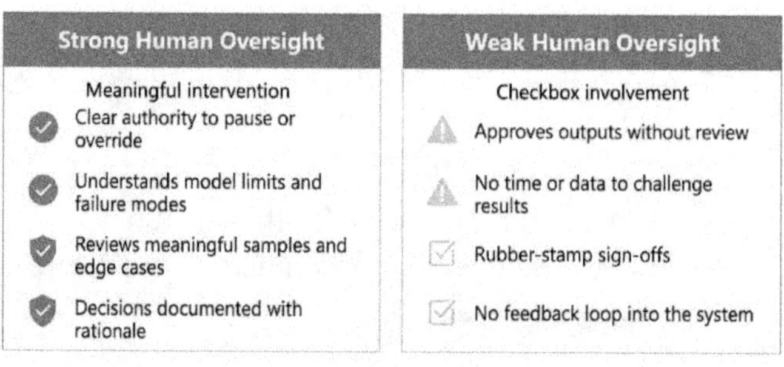

12.7 Designing human oversight into workflows

To embed real oversight, you must design it explicitly into your workflows:

- Identify high-impact decisions where AI should **recommend**, not decide.
- Define thresholds (confidence, risk level, exception patterns) that trigger human review.
- Provide UIs that highlight uncertainty and relevant context, not just scores.
- Reduce cognitive load by summarizing key factors, not dumping raw data.

Human oversight design checklist

- High-impact decisions have defined human checkpoints.
- Criteria for when AI can act autonomously vs when a human must review are documented.
- Interfaces support easy review and override (with logging).
- Oversight responsibilities are part of job descriptions and capacity planning.

Human-in-the-Loop Workflow

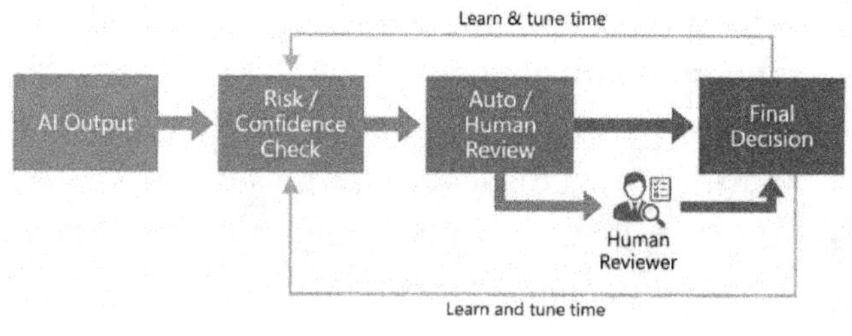

Template: Human oversight specification (per system)

12.8 Human Oversight Specification – Template

For system:

1. Decisions and risk levels

- List decisions this system influences or makes.
- Classify decisions by impact (Low / Medium / High).

2. Oversight rules

- For each decision type:
    - When can the AI decide autonomously?
    - When must a human review before action?
    - When must a human be able to override after the fact?

3. Interface & information

- What information is shown to reviewers (inputs, explanations, prior history)?
- How is uncertainty communicated?

4. Responsibilities & capacity

- Who performs oversight (roles/teams)?
- How is time allocated and performance measured?

5. Logging & learning

- How are human overrides logged and analyzed?
- How do oversight outcomes feed back into model and workflow improvements?

Psychological safety: making it safe to raise red flags

You can have checklists and oversight on paper. They will not be used if people fear punishment for slowing things down or questioning popular projects. Psychological safety means:

- People can say, "I am not comfortable with this," without being labeled as blockers.
- Failing fast on a risky idea is seen as learning, not failure.
- Leaders publicly support and reward teams that raise tough issues.

As a manager, you set the tone by how you respond the first few times someone raises a concern. If you react defensively or rush past it, others will watch and stay quiet.

12.9 Psychological safety signals

- People volunteer risk concerns in meetings, not just in private.
- Junior staff sometimes challenge assumptions—and are taken seriously.
- Incident reviews focus on systems and processes, not blame games.

"Psychological Safety Loop"

Speak Up

Taken Seriously

Action / Feedback

More Willingness to Speak Up

Template: "Raise a Flag" lightweight reporting channel

Create a simple, low-friction mechanism for raising concerns:

Raise a Flag – Template

244

- System or project name:
- What are you seeing or worried about? (short description)
- Category (if known): [Fairness] [Safety] [Misuse] [Privacy] [Compliance] [Other]
- How severe might the impact be? [Low / Medium / High / Not sure]
- Any examples or evidence?
- Optional: your name and contact (can be anonymous if needed).

Behind the scenes, ensure:

- Someone is responsible for triaging flags.
- There is a defined SLA for responding.
- Aggregated themes are reported to governance forums.

Case pattern: a high-stakes oversight culture

Imagine a healthcare organization using AI to support diagnostic suggestions. They embed:

- Required training for clinicians on AI strengths and limitations.
- Interfaces that require clinicians to confirm or reject AI suggestions, with reasons.
- Regular "Grand Rounds" sessions where AI-influenced cases are reviewed for learning.

- A culture where disagreeing with the AI and explaining why is normal, not rebellious.

This environment does not eliminate errors, but it **catches** and **learns from** them faster—and keeps the clinician, not the model, at the moral center of decision-making.

Case pattern: a public sector team building ethical reflexes

Consider a public sector agency using AI to prioritize benefit reviews. They:

- Run cross-functional workshops where caseworkers, data scientists, and policy staff walk through hypothetical scenarios and identify fairness risks.
- Include caseworkers in design decisions so frontline realities shape thresholds and workflows.
- Normalize "pausing" an AI feature when fairness concerns arise, with executive support.
- Make fairness and recourse metrics part of leadership dashboards.

Here, culture is reinforced through **shared problem-solving** and visible leadership backing for responsible pauses.

Ethical Reflexes in Public Sector Teams

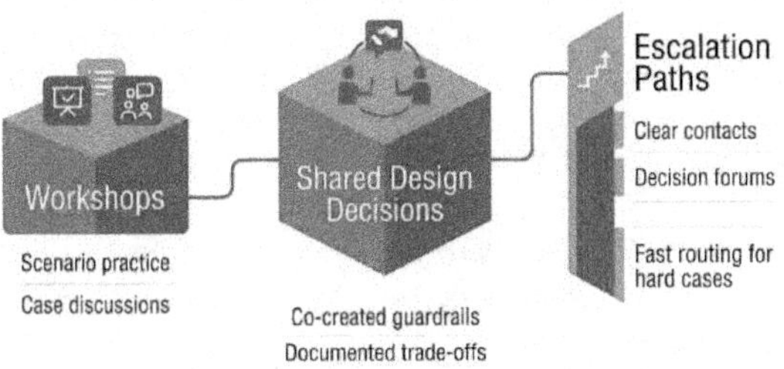

The Manager's Culture & Oversight Playbook (90-day plan)

To strengthen culture, training, and oversight:

0–30 days: Diagnose and listen

- Ask teams where they feel least comfortable raising AI concerns.
- Review recent incidents and retrospectives for cultural patterns (silence, rushed decisions).
- Inventory current training and oversight mechanisms.

31–60 days: Build tools & training

- Develop the Responsible AI Design & Review Checklist and Human Oversight Specification templates.

- Create role-specific training sessions with concrete scenarios.
- Launch a simple "Raise a Flag" channel with a clear process behind it.

61–90 days: Model and reward behavior

- Use live projects to pilot the checklists and oversight templates.
- Publicly acknowledge teams that raise and address difficult issues.
- Integrate responsibility behaviors into performance and recognition conversations.

Culture & Oversight Roadmap

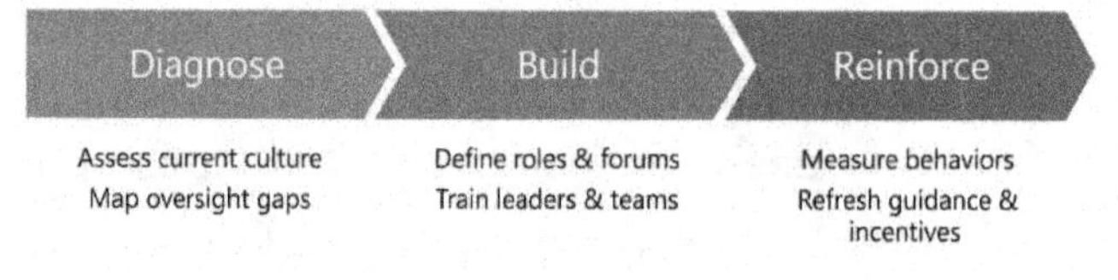

12.10 The AI Reality Check: culture is the multiplier

Culture is not a soft add-on; it is the multiplier that makes all your technical and governance investments either effective or fragile. With the wrong culture:

- Checklists are ignored.
- Oversight is performative.
- Incidents repeat because root causes are social, not technical.

With the right culture:

- Small issues surface before they become big ones.
- People use tools and templates as intended.
- Governance feels like a shared practice, not an external imposition.

Culture as Multiplier

The Monday Morning Test

On Monday morning, a serious AI manager should be able to answer:

1. Do my teams know what "responsible AI" means here, in practical, everyday terms?

2. When was the last time someone raised a concern about an AI system, and what happened afterward?
3. Where, in our workflows, is human oversight clearly designed and resourced (not just mentioned)?
4. What decision checklists or templates are actually being used by teams—not just existing in a repository?
5. How am I personally reinforcing a culture where speaking up about AI risks is expected, supported, and acted upon?

If some answers are "I am not sure," that is your leadership agenda. Technology may enable AI, but culture, training, and human oversight determine whether it helps your mission—or quietly undermines it.

13 Measuring and Reporting AI Impact

Opening vignette

Your AI modernization program is in full swing. Models are in production, governance forums meet regularly, and teams are busy. A board member asks you: "Are our AI systems making us safer and fairer—or just faster?" The regulator wants evidence of how you monitor risk. Internal stakeholders want to know if AI is actually improving outcomes for customers and staff. You realize your dashboards show model performance and cost savings, but not **the impact on responsible AI**. Governance exists, but its effectiveness is largely a matter of opinion. In that moment, it is clear: to lead AI responsibly, you must **measure and report** its impact—not just on efficiency, but on ethics, safety, fairness, and compliance.

13.1 Why governance needs metrics

Governance without metrics drifts into theater. People talk about principles and frameworks, but cannot tell whether risk is going up or down, or which interventions work. Metrics anchor your governance in observable reality.

For IT managers, responsible AI metrics:

- Turn vague concerns into concrete, trackable issues.
- Enable prioritization of remediation efforts.
- Provide evidence to boards, regulators, and the public that you are managing AI deliberately.

Without them, you rely on anecdotes, isolated incidents, and individual judgment. With them, you can steer.

What to measure: the four dimensions of responsible AI impact

To avoid metric overload, focus on four core dimensions:

1. **Ethics & fairness:** Are outcomes aligned with your values and fair across groups?
2. **Safety & reliability:** Are you avoiding harm and managing failure modes?
3. **Compliance & governance:** Are you living up to your policies and obligations?
4. **Value & adoption:** Are you responsibly achieving business/mission value?

Each dimension should have a small, curated set of Key Performance Indicators (KPIs) that you track over time.

Four Dimensions of Responsible AI KPIs

Ethics / Fairness	Safety / Reliability
bias rate disparities complaints on unfair outcomes	incident count / severity model uptime & failure rate
Compliance / Governance	Value / Adoption
regulatory findings policy exceptions & overdue actions	active users & usage depth business value delivered customer value creation

13.2 Designing responsible AI KPIs: principles

Good, responsible AI KPIs are:

- **Relevant:** Tied to real risks and goals (e.g., fairness in lending, safety in clinical support).
- **Actionable:** When a metric moves, you know who should act and how.
- **Balanced:** Not just performance or cost; also fairness, safety, and compliance.
- **Feasible:** Data is available and reliable enough to track consistently.

Avoid vanity metrics that look sophisticated but never influence decisions.

13.3 Ethics and fairness KPIs

Ethics and fairness KPIs translate abstract values into measurable patterns. Examples:

- **Outcome parity:** Differences in approval rates, benefit assignment, or prioritization across groups (where appropriate).
- **Error parity:** Differences in false positive/negative rates across groups.
- **Review rates:** Percentage of cases from certain segments that need human review due to uncertainty or conflict.
- **Complaint rates:** Fairness-related complaints per 1,000 decisions, by group.

13.4 Ethics & fairness KPI checklist (per system)

- Relevant groups identified (demographic, geographic, segment).
- At least one outcome parity metric is tracked where meaningful.
- Error parity or quality metrics monitored across key groups.
- Fairness-related complaints and flags are logged and categorized.

Example Fairness KPI Dashboard

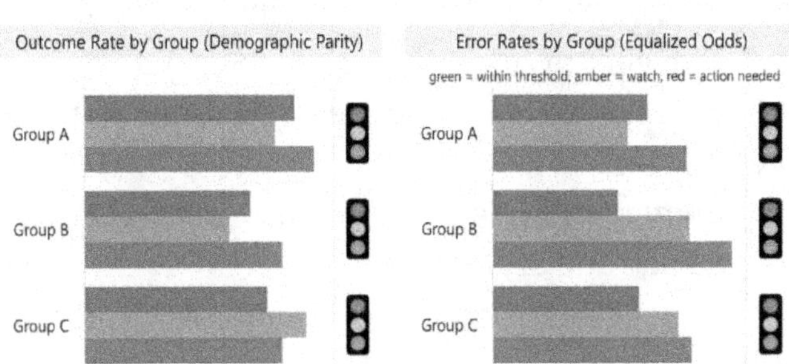

Safety and reliability KPIs

Safety and reliability metrics tell you whether your AI behaves acceptably in real-world conditions. Examples:

- **Incident frequency:** Number and severity of AI-related safety incidents (e.g., harmful content, incorrect high-impact recommendations) per period.
- **Hallucination rate (for generative systems):** Percentage of sampled outputs that fail factual or policy checks.
- **Override rate:** Percentage of AI decisions overridden by human reviewers (by reason).
- **Degradation indicators:** Changes in key performance metrics over time (e.g., accuracy drift).

Safety & reliability KPI checklist

- AI-related incidents are clearly defined and consistently logged.
- Incident counts and severities monitored and trended.
- Human override rates tracked and analyzed (by reason and segment).
- Model performance drift indicators monitored and thresholded.

13.5 Compliance and governance KPIs

Compliance and governance KPIs show whether your governance system is functioning, not just existing on paper. Examples:

- **Coverage:** Percentage of AI systems with complete documentation (model cards, fact sheets, risk assessments).
- **Review freshness:** Percentage of high-risk systems with a governance review in the last X months.
- **Control implementation:** Percentage of required controls implemented by risk tier.

- **Audit findings:** Number and severity of internal/external audit findings related to AI per year.

Compliance & governance KPI checklist

- All AI systems are cataloged and classified by risk.
- Documentation completeness is measured and reported.
- Review cadence compliance is tracked (especially for high-risk systems).
- Audit findings and remediation status are monitored.

13.6 Value and adoption KPIs (with a responsibility lens)

You still need traditional impact metrics—cost savings, productivity gains, and error reduction—but they should be interpreted through a responsibility lens.

Examples:

- **Value metrics:** Processing time reduction, increased throughput, error reduction, revenue uplift.
- **Adoption metrics:** Percentage of target users actively using AI features.

- **Balanced metrics:** Value metrics contextualized by fairness and safety (e.g., "Processing time down 30% with no increase in fairness complaints or safety incidents").

Value & adoption KPI checklist

- Clear value metrics defined per system and tracked consistently.
- Adoption metrics segmented by user group (to detect uneven impacts).
- Value metrics reviewed alongside ethics/safety/compliance KPIs, not in isolation.

13.7 Building an AI impact scorecard (system level)

To make metrics usable, consolidate them into a simple **impact scorecard** for each major AI system.

AI System Impact Scorecard – Template

For system:

1. Summary

- Purpose and domain:
- Risk tier:

2. Ethics & fairness

- Key fairness metrics and latest values:
- Fairness flags/complaints (recent period):

3. Safety & reliability

- Incident summary (count, severity):
- Override rates and reasons:
- Performance drift indicators:

4. Compliance & governance

- Documentation completeness:
- Last governance review date and outcome:
- Open audit findings (if any):

5. Value & adoption

- Key value metrics (e.g., processing time, error rates, savings):
- Adoption metrics (usage, user groups):

6. Overall assessment

- Concise narrative (1–2 paragraphs) summarizing impact and risk posture.

13.8 Portfolio-level AI impact reporting

Beyond individual systems, leaders want a **portfolio view** of AI impact. Your portfolio-level reporting might include:

- Number of AI systems in production, by domain and risk tier.
- Aggregate value metrics (e.g., annualized hours saved, error reductions) with context.
- Aggregate ethics, safety, and compliance metrics (e.g., fairness flags, incidents, documentation coverage).
- Trends in key risk metrics (e.g., incidents, audit findings, high-risk systems without recent review).

This gives boards and executives a concise picture: where AI is helping, where risk is concentrated, and whether controls are improving over time.

Portfolio AI Impact Dashboard

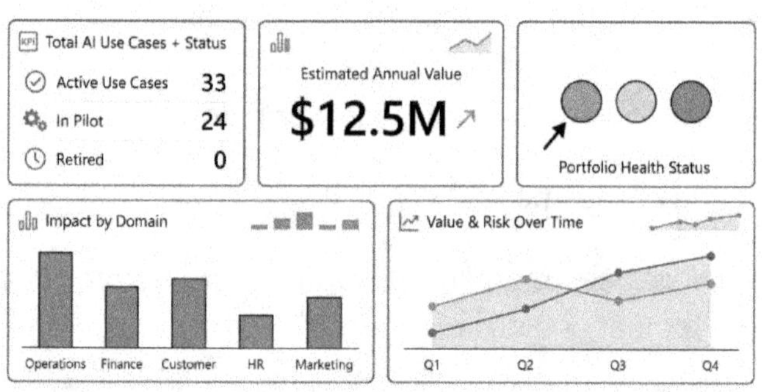

Template: Quarterly AI impact report (management/board)

Quarterly AI Impact Report – Template

1. Executive summary

- High-level overview of AI deployments, value delivered, and risk posture.

2. AI footprint

- Number of AI systems in production (by domain and risk tier).
- Notable new deployments or retirements this quarter.

3. Value & performance

- Key aggregate metrics (e.g., time saved, throughput, error reduction).
- Highlights of important success stories or improvements.

4. Ethics & fairness

- Summary of fairness KPIs across critical systems.
- Notable fairness-related issues and actions taken.

5. Safety & reliability

- Incident summary (counts by severity and type).
- Patterns in overrides, degradations, and mitigations.

6. Compliance & governance

- Documentation and review coverage metrics.
- Audit activities and key findings/remediations.

7. Forward look

- Planned enhancements, upcoming high-impact deployments, and risk hotspots to watch.

13.9 External reporting: regulators and the public

For some organizations and use cases, you will need **external-facing reporting**:

- Regulators may request evidence of risk management and outcomes for high-risk AI systems.
- Customers or citizens may expect transparency reports on how AI affects them.
- Partners and investors may want assurance that your AI practices are credible and futureproof.

External reporting should:

- Use plain language and avoid technical jargon.
- Focus on key impacts, not all internal details.
- Be honest about limitations and ongoing improvements.

13.10 Assurance audits: turning scrutiny into a trust asset

Assurance audits—internal or third-party—are becoming a key mechanism for validating AI practices. They can cover:

- Model performance and testing.
- Fairness and safety controls.
- Governance processes and documentation.
- Compliance with internal policies and external expectations.

Handled well, audits are not threats; they are **credibility tools**. You can reference them in board discussions, regulatory engagements, and public communications.

Assurance audit readiness checklist

- Scope and objectives for AI audits defined (internal/external).
- Key artifacts (policies, risk assessments, documentation, logs) organized and accessible.
- Owners identified for responding to findings.
- Audit findings tracked with remediation plans and timelines.

Template: AI assurance audit summary (per audit)

AI Assurance Audit Summary – Template

- Audit type (internal/external):
- Scope (systems, domains, time period):
- Objectives (what was assessed):

1. Key findings

- Summary of major strengths.
- Summary of significant issues.

2. Detailed findings

- Finding ID, description, severity, and affected systems.

3. Remediation

- Actions planned, owners, target dates.

4. Lessons learned

- Process or governance improvements identified.

5. Follow-up

- Planned re-assessment dates or triggers.

Making metrics "live": governance using numbers, not anecdotes

Metrics only matter if they are **used**. To make your AI impact metrics live:

- Build them into regular governance meetings (AI Review Board, risk committees, steering groups).
- Use them to prioritize work (e.g., which systems get remediation resources, where to slow or halt expansion).
- Track decisions back to metrics: "We tightened thresholds because incidents were rising," "We paused rollout due to fairness disparity."

This creates a virtuous cycle: metrics inform decisions, decisions change behavior, and behavior changes metrics.

The Manager's Measurement & Reporting Playbook (90-day plan)

To operationalize AI impact measurement and reporting:

0–30 days: Define and select

- Identify your top 3–5 AI systems by impact and risk.
- Define minimally responsible AI KPIs for each (ethics/fairness, safety, compliance, value).
- Agree on common definitions and data sources for those KPIs.

31–60 days: Build scorecards and dashboards

- Create system-level AI Impact Scorecards for the selected systems.

- Build a simple portfolio-level dashboard aggregating key indicators.
- Establish who owns each metric and how often it is updated.

61–90 days: Embed in governance & reporting

- Use the scorecards and portfolio dashboard in at least one governance forum (e.g., AI Review Board, risk committee).
- Produce a lightweight quarterly AI impact report for leadership.
- Adjust metrics and formats based on feedback, focusing on clarity and actionability.

The AI Reality Check: avoiding "KPI theater."

Just as there is governance theater, there can be KPI theater: impressive-looking dashboards that do not change decisions. Signs include:

- Too many metrics; nobody can say which matters most.
- Metrics are presented without owners or actions.
- Metrics focused only on performance and cost, ignoring fairness and safety.
- Reports produced for their own sake, not tied to governance or strategy discussions.

Real measurement and reporting become visible when:

- Metrics are debated in meetings and drive trade-off decisions.
- Some initiatives slow down or change because risk metrics demand it.
- Leadership can answer "Are we managing AI risk better this quarter than last?" with evidence, not anecdotes.

The Monday Morning Test

On Monday morning, a serious AI manager should be able to answer:

1. For my most important AI systems, what are the 5–10 key KPIs that describe their ethical, safety, compliance, and value impact?
2. Where do those metrics live (which dashboards, which reports), and who keeps them current?
3. How often do governance forums review these metrics—and can I point to a recent decision influenced by them?
4. What does our current AI impact summary say to the board or leadership about risk and value—would it withstand external scrutiny?
5. If a regulator or critical stakeholder asked for evidence that we measure and manage AI responsibly, what would I show them this week?

If the answers are thin or fragmented, that is your measurement agenda. Metrics are not about perfection; they are about turning Responsible AI from a belief into a **trackable performance discipline** that you can explain, defend, and improve.

---

14 From Principles to Practice: The Responsible AI Roadmap

Opening vignette

You have put in the work. You have chapters' worth of principles: ethics, safety, fairness, transparency, accountability, governance, risk, policy, data, culture, and metrics. You have run a few pilots, created some templates, and held your first AI Review Board meetings. However, as you look across your portfolio, you see fragmentation: some teams use the new tools, some still "go fast and ask later," and leadership keeps asking, "Where are we on Responsible AI as an organization?" You realize you do not just need more concepts—you need a **roadmap**: a clear sequence of steps that can move your organization from scattered efforts to a coherent Responsible AI operating model, without stalling modernization.

14.1 Why you need a roadmap, not more principles

Principles without a roadmap create frustration. Teams know "what good looks like," but not how to get from their current reality to that future state. A roadmap:

- Breaks Responsible AI into **phases** with concrete outcomes.
- Lets you **sequence investments** instead of trying to "do everything now."
- Gives leadership a **visible plan** to support and hold you accountable to.

In practice, the roadmap is your bridge between this book and "what we do over the next 12–24 months."

Principles → Roadmap → Practice

14.2 The four-phase Responsible AI roadmap

A pragmatic roadmap for IT managers leading AI modernization has four phases:

1. **Assess:** Understand where you are—systems, risks, gaps, and strengths.
2. **Pilot:** Apply Responsible AI practices deeply in a small set of high-leverage projects.
3. **Scale:** Standardize and roll out governance, tools, and templates across the portfolio.
4. **Improve:** Continuously refine based on metrics, incidents, and changing context.

You do not need to move every part of the organization at the same pace, but you do need to know which phase you are in for each domain and system.

Phase 1: Assess – get a clear picture of your starting point

Assessment is about replacing assumptions with facts. Before you can lead, you must see your AI landscape clearly.

Key tasks in Assess:

- **Inventory AI systems:** What is in production, in pilot, and in the pipeline? Who owns each system?
- **Classify risk:** Assign risk tiers based on impact and domain (e.g., high, medium, low).
- **Map governance and data:** Identify which systems have documentation, risk assessments, and clear data owners—and which do not.

- **Identify gaps:** Where are you most exposed (e.g., high-risk systems with weak controls, no oversight, or poor data stewardship)?

14.3 Assess phase checklist

- AI system inventory created and validated with domains.
- Risk tier assigned to each system.
- Ownership documented (business, technical, data, risk).
- Current controls (governance, data, monitoring) mapped.
- Top gaps and exposure areas identified and prioritized.

14.4 Template: Responsible AI baseline assessment summary

Use this as the anchor document for Phase 1:

Responsible AI Baseline Assessment – Template

1. Scope

- Business units/domains covered:
- Timeframe of assessment:

2. AI footprint summary

- Number of AI systems in production (by domain and risk tier):
- Key high-impact systems identified:

3. Governance & controls status

- Percentage with model cards and fact sheets:
- Percentage with risk assessments and documented owners:
- Presence of AI governance forums (Yes/No, brief description):

4. Data and monitoring status

- Critical datasets identified (Yes/No, partial):
- Data quality/lineage coverage for high-risk systems:
- Monitoring/metrics coverage (ethics, safety, compliance):

5. Key gaps and risks

- Top 5 gaps with brief descriptions and potential impact:

6. Recommended focus areas for next 6–12 months

- Priority domains or systems:
- Priority governance, data, or culture improvements:

Phase 2: Pilot – go deep on a few high-leverage systems

Pilots move Responsible AI from theory to practice. Choose 2–5 systems that:

- Are high-impact or high-risk.
- Have engaged owners willing to experiment.
- Represent different domains or patterns you care about.

Apply the full Responsible AI toolkit to these pilots:

- Risk assessment and register.
- Ethics, fairness, and safety checklists.
- Model cards, fact sheets, and governance reviews.
- Data integrity and stewardship practices.
- Human oversight design and training.
- Metrics and impact scorecards.

Pilot phase checklist

- Pilot systems are selected with clear criteria and sponsorship.
- Full Responsible AI lifecycle applied (from intake to monitoring).
- Issues and friction points documented (what was hard, what duplicated effort).
- Early wins captured (risk avoided, clarity gained, better decisions made).

Pilot Deep Dive

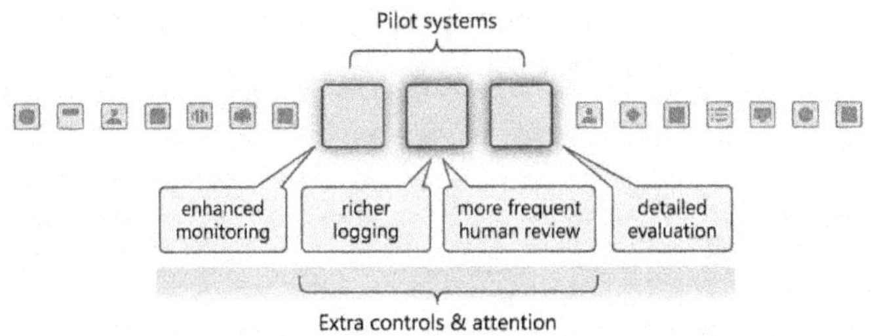

14.5 Template: Pilot project Responsible AI plan

Pilot Responsible AI Plan – Template

For each pilot system:

1. System overview

- Name, domain, risk tier:
- Owners (business, technical, data, risk):

2. Scope of Responsible AI practices

- Which artifacts will be created/updated (model card, fact sheet, risk assessment, etc.):
- Which checklists will be applied (fairness, safety, oversight, data, policy):

3. Governance

- What forums will review this system (and when):
- Key decision points (go/no-go, thresholds, risk acceptance):

4. Metrics & monitoring

- Responsible AI KPIs to track (fairness, safety, compliance, value):
- Initial thresholds and alerting rules:

5. Learning goals

- What you want to learn about applying Responsible AI in this organization:
- How insights will be captured and shared:

Phase 3: Scale – standardize and embed across the portfolio

Once pilots show what works (and what doesn't), you move into scaling. Scaling is not "copy everything blindly"; it is about **standardizing what proved valuable and right-sizing it** for the rest of your portfolio.

Key moves in Scale:

- **Standard templates:** Model cards, fact sheets, risk assessments, accountability sheets, Responsible AI design briefs, test add-ons, etc.

- **Standard processes:** AI intake, risk classification, governance workflow, data stewardship routines, and monitoring cadences.
- **Tooling integration:** Embedding checklists and documentation into existing platforms (ticketing, CI/CD, MLOps, data catalogs).
- **Training programs:** Role-based training rolled out beyond pilot teams.

Scale phase checklist

- Core templates and checklists refined from pilots and published as standards.
- AI governance workflow documented and socialized.
- AI and data governance roles defined across domains.
- Top-priority systems beyond pilots scheduled for onboarding to new practices.

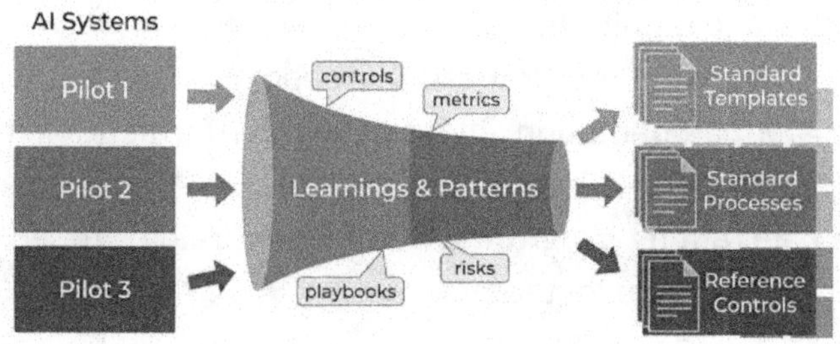

Template: Responsible AI rollout plan (scale phase)

Responsible AI Rollout Plan – Template

1. Scope and prioritization

- Domains included in this rollout:
- System groups and sequence (e.g., high-risk first, then medium):

2. Standards and tools

- Templates to be mandated (list):
- Processes to be adopted (intake, reviews, monitoring):
- Tool changes required (forms, workflows, dashboards):

3. Timeline and milestones

- Milestone 1: Standards approved and communicated.
- Milestone 2: High-risk systems onboarded.
- Milestone 3: Medium-risk systems onboarded.
- Milestone 4: Portfolio-level metrics operational.

4. Support and enablement

- Training and office hours:
- Help channels and points of contact:

5. Risks & mitigations

- Potential resistance points (e.g., perceived overhead):
- Mitigation strategies (simplification, phased adoption, executive sponsorship):

Phase 4: Improve – make Responsible AI a continuous discipline

Responsible AI is not "one and done." New use cases, regulations, vendors, and incidents will continue to reshape your landscape. The Improve phase is about building the **muscle of continuous improvement**:

- Regularly reviewing metrics and incidents to refine controls and processes.
- Updating templates and workflows based on user feedback.

- Adjusting risk tiers and controls as your portfolio and risk appetite evolve.
- Conducting periodic assurance audits and maturity assessments.

Improve phase checklist

- Quarterly AI risk and impact reviews are in place.
- Templates and checklists have owners and version history.
- Lessons learned from incidents and audits feed into process updates.
- Roadmap is revisited annually to reflect new realities.

Continuous Improvement Loop

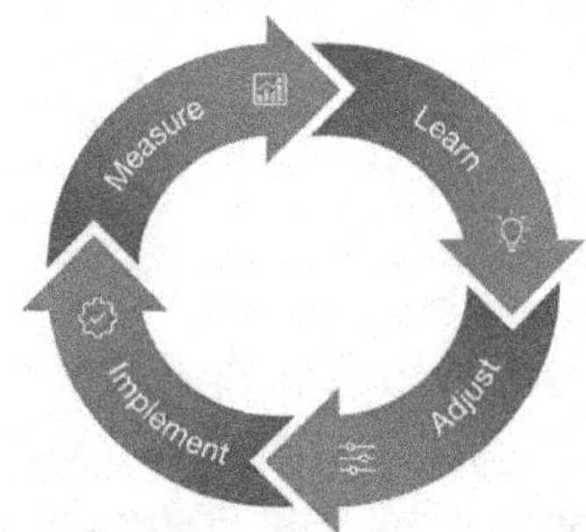

14.6 Connecting the roadmap to business outcomes

Responsible AI is not a side project; it must be tied to **business and mission outcomes**. At each phase, explicitly link Responsible AI actions to outcomes leaders care about:

- **Risk reduction:** Fewer incidents, better regulatory posture, reduced remediation costs.
- **Trust and adoption:** Higher uptake of AI tools by staff and customers, fewer complaints.
- **Operational clarity:** Faster approvals, fewer surprises, and less firefighting.
- **Strategic flexibility:** Ability to scale AI into sensitive domains because governance is strong.

You can use your metrics (from Chapter 13) to show these outcomes in numbers rather than just narratives.

Template: Responsible AI roadmap one-pager (for leadership)

Responsible AI Roadmap – Leadership One-Pager Template

1. Vision

- One sentence on what Responsible AI means for your organization.

2. Phases and timelines

- Assess: [Dates, main activities]
- Pilot: [Dates, main activities]
- Scale: [Dates, main activities]
- Improve: [Ongoing, main activities]

3. Key focus areas

- Governance & risk
- Data & stewardship
- Culture & human oversight
- Metrics & reporting

4. Expected outcomes

- Risk reduction goals:
- Trust and adoption goals:
- Operational efficiency goals:

5. Ask leadership

- Sponsorship needed:
- Decisions or resources required:

14.7 Sequencing: which levers to pull first

You cannot do everything at once. A sensible sequencing for most organizations:

1. **Visibility:** Inventory systems and set basic risk tiers.

2. **High-risk focus:** Apply full Responsible AI treatment to 2–5 high-impact systems.
3. **Governance backbone:** Stand up or refine AI governance forums and workflows.
4. **Data foundations:** Identify critical datasets and basic data stewardship.
5. **Culture & training:** Launch targeted training and "raise a flag" mechanisms.
6. **Metrics:** Start simple, system-level impact scorecards for key systems, then a portfolio view.

This sequence ensures you do not overwhelm teams and that early wins build credibility.

14.8 The Manager's Integrated Roadmap Checklist

This checklist ties the whole book together into a single "are we set up to execute?" view:

Integrated Responsible AI Roadmap Checklist

A. Foundations

- Systems inventory and risk tiers completed.
- AI and data governance roles defined (owners, stewards, boards).

B. Governance & risk

- Governance operating model documented (workflow, forums, decision rights).
- Risk framework and templates (assessments, registers) in use for high-risk systems.

C. Responsible lifecycle

- Responsibility by Design checkpoints embedded in intake, design, build, deploy, and monitor.
- Checklists and templates are used in real projects, not just published.

D. Data & culture

- Critical datasets identified and stewarded; basic quality/lineage in place.
- Training programs launched for engineers, product owners, and frontline staff.
- Human oversight is designed and resourced for high-impact decisions.

E. Metrics & reporting

- System-level impact scorecards exist for key AI systems.
- Portfolio-level AI impact reporting started (even if simple).
- Metrics are discussed in governance forums and drive decisions.

14.9 Monday Morning: your first 90 days from this book

To make this chapter immediately useful, here is a concrete 90-day plan you can start on Monday.

Days 1–30: See clearly

- Build or refine your AI system inventory and risk tiers.
- Identify top 3–5 high-impact/high-risk systems.
- Draft a baseline Responsible AI assessment summary.

Days 31–60: Prove it on pilots

- For those systems, apply the full Responsible AI toolkit (risk, governance, data, culture, metrics).
- Document what works, what is heavy, and what is missing.
- Prepare system-level impact scorecards and share with leadership.

Days 61–90: Stabilize and communicate

- Draft your Responsible AI roadmap one-pager with phases, focus areas, and outcomes.
- Define your scale and improve strategy (which standards will be universal, which processes will be mandated).

- Agree with leadership on 2–3 clear Responsible AI goals for the next 12 months.

14.10 The final Monday Morning Test

By the end of this book, a serious AI manager should be able to sit down on Monday morning and answer:

1. Do we have a clear, shared view of where AI is in our organization and where the biggest risks are?
2. Can I name the 3–5 AI systems that will be our proving ground for Responsible AI practices in the next quarter?
3. Do we have an agreed governance operating model, and are we piloting it on real projects?
4. Are our data, culture, and metrics workstreams anchored to those pilots, not drifting separately?
5. Can I show my leadership team a one-page roadmap that explains how we will move from principles to practice over the next 12–24 months?

If you can answer "yes" to most of these, you are not just talking about Responsible AI—you are leading it. If some answers are still "not yet," you now have a roadmap to get there, step by step, with clarity, discipline, and a culture that knows how to make AI serve your mission, not the other way around.

Claude Louis-Charles PhD

15 Responsible AI and Governance

15.1 Contextualizing AI Governance

Academic, policy, and industry interest in various aspects of Artificial Intelligence (AI) continues to grow exponentially. This is because of the technology's unprecedented impact across sectors such as education, healthcare, finance, and public administration. However, much as AI's benefits to humanity are well-documented, its potential for harm is equally recognized. Stakeholders appreciate that AI can lead to unintended, unforeseeable, and injurious outcomes, posing new risks that governments and other regulators are ill-equipped to handle (Taeihagh, 2021). Crucially, the technology's poorly understood nature casts doubt on the efficacy of traditional governance tools such as taxes, legislation, and subsidies. At the very least, preexisting governance mechanisms need to be constantly re-imagined and revamped. Relatedly, the speed and scale of AI adoption threaten to outpace regulatory responses (Taeihagh, 2021). Indeed, policymakers are playing catch-up with AI applications in high-risk areas such as transport and healthcare. Unless robust governance practices and tools are developed and applied to minimize the adverse effects of AI, human interests, including the fundamental pillars of civilization, will be endangered. Consequently, this paper analyzes the impacts of AI on humans and society, delineates the principles of responsible AI, and summarizes the regulatory framework for AI. Further, it addresses how responsible AI governance and risk management can be implemented, as well as the operationalization of AI projects and systems. In the end, it concludes with a glimpse of ongoing AI issues and concerns.

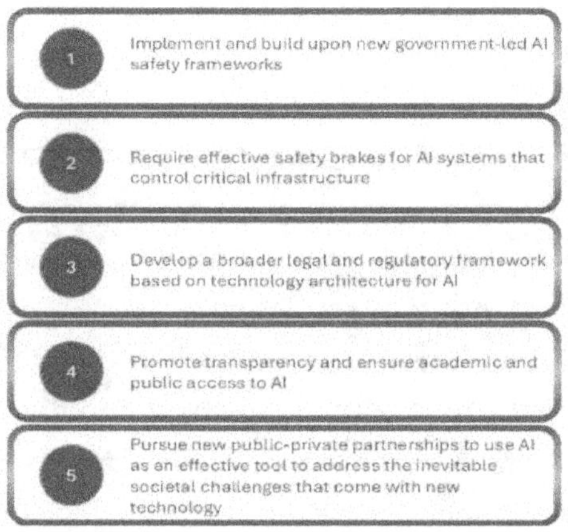

Figure 2 Five points to AI Governance

15.2 AI Impacts: Positives, Perils, and Risks

Robust AI governance can only be designed and implemented with an intimate understanding of its impacts. These include positive impacts that support its growth and adoption, perils apparent from current and projected AI uses, and risks that may be imminent in its deployment now and in the future. A comprehensive assessment of all impacts is a long shot, but a useful overview is provided herein.

15.2.1 Positive Impacts

AI has had positive effects on industrial efficiency through automation and intelligent machines. Automation has replaced humans in many blue-collar industries due to its ability to sustain productivity over long periods with high accuracy (Shaukat et al., 2020). Unlike humans, who become fatigued and erratic after extended periods of work, machines can maintain efficiency indefinitely. Automating mundane tasks also allows humans to focus on other activities that require creativity and human judgment. For instance, Ali et al. (2023) observe that automation in the healthcare industry has enabled medical professionals to focus on more fundamental tasks related to patient management and healthcare resource management. Automating such tasks that would otherwise unnecessarily expend the time and energy of medical personnel frees medical professionals to focus on more meaningful tasks. Still on efficiency, automation and intelligent robots can perform tasks that are dangerous to humans in factories and other industrial settings (Shaukat et al., 2020). For instance, robots capable of performing hazardous tasks in the agricultural and mining sectors are experiencing rising demand. Automation and intelligent machines are cheaper, faster, and more accurate in handling repetitive tasks in different industries, resulting in improved efficiency.

In the healthcare industry (Cardiology Example), AI has revolutionized disease detection, diagnosis, and monitoring. AI applications can gather, analyze, and generate predictions using various types of medical data, such as Electronic Health Records (EHRs) and imaging data (Ali et al., 2023). Human beings cannot capture and manipulate the data at the speed, with the diligence, and with the accuracy that AI tools provide. More importantly, AI applications are offering critical decision-making support to healthcare professionals in disease detection, diagnosis, and prevention. For instance, AI diagnostic algorithms are used in the early detection of breast cancer (Lee & Yoon, 2021). This can be achieved in many ways, including offering a "second opinion" on the interpretation of radiological images. At the same time, Internet of Things (IoT) applications use AI to monitor patients' vital signs post-discharge, resulting in lower readmission rates and fewer emergency department visits (Lee & Yoon, 2021). These are only a small fraction of the revolutionary impacts that AI is delivering in the healthcare space. Besides, more intriguing benefits are expected as AI continues to evolve and gain broader adoption across different healthcare contexts.

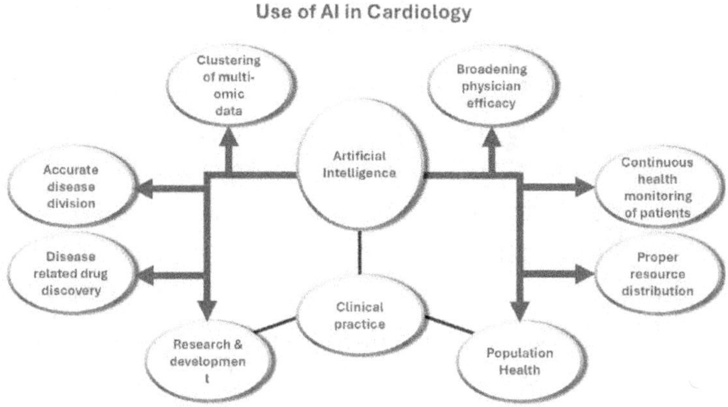

Figure 3 AI in Cardiology

In the public sector, AI applications have been widely credited for ushering in a new era of administrative efficiency. Case studies from around the world today show how transformative AI has been in improving public services. For instance, AI has been used to identify and divert vulnerable people who cannot pay penalties from enforcement action in New South Wales (NSW) (Alhosani & Alhashmi, 2024). This allows the revenue department to offer other settlement options to such individuals. In the past, revenue officials could not determine that an individual was vulnerable until enforcement, which exacerbated injustices against such disadvantaged people. The French government has also turned to AI to improve building annotations. The said government uses an AI application to locate previously unidentified buildings, a new practice that reportedly led to an additional €10 million in tax revenues (Alhosani &

Alhashmi, 2024). The use of such AI applications in physical planning and enforcement is also evident in countries like Australia and the US. Notably, the US government uses an AI application called Near Map to monitor physical infrastructure, thereby identifying tax cheats and illegal property transfers (Alhosani & Alhashmi, 2024). These tasks would have required a substantial investment in material and human resources in the absence of AI. The above applications demonstrate the transformative potential of AI in the public sector.

AI has revolutionized many other products and services across sectors and holds positive long-term prospects. Indeed, an exhaustive review of all positive impacts flowing from the technology is inconceivable. As Peter (2021) explains, AI is implicated in virtual assistants such as Siri, smart cities, smart homes, driverless cars, autonomous weapons, and many other applications. Some routine smartphone applications, such as Google Translate, also use AI. The applications are so pervasive that some may easily go unrecognized by users. For instance, AI is writing poems and novels, driving targeted advertising, helping to eliminate disinformation and pornography on social media, trading stocks, and assigning credit scores (Peter, 2021). Users interact with such AI functionality daily without distinguishing it from human activities. In the long term, AI's impact will be undoubtedly profound. For instance, AI efficiencies are expected to lower costs across many industries. Ali et al. (2023) estimate that AI implementation in the healthcare sector will reduce annual spending by US$ 150 billion by 2026. This will help to resolve the problem of burgeoning healthcare spending that has persisted in the US healthcare system. By 2030, AI is estimated to add US$16 trillion to the global economy (Peter, 2021). Thus, AI has been remarkably impactful across sectors and will continue to yield positive effects as we advance. Cumulatively, the said benefits explain why AI adoption will only increase in speed and scale.

15.2.2   Perils and Risks

Current and projected benefits of AI must, however, be carefully balanced against its downsides and risks. Invariably, any AI application that offers benefits also comes with downsides or risks. For instance, while much is made of AI's capacity to learn from data independently, such learning takes the technology outside the framework of programmed rules, which might result in unanticipated outcomes (Taeihagh, 2021). Some of the unexpected outcomes of AI may be negative or even catastrophic. Equally, trained algorithms may fail to adapt to new contexts, leading to highly undesirable results. The conditions under which an algorithm was trained are not the same as those of the natural environment in which it will be operationalized, which might lead to dangerous outcomes. For instance, fatal crashes involving Tesla's semi-autonomous vehicles have been reported due to the algorithm's misunderstanding of unusual environmental conditions not encountered during training (Taeihagh, 2021). Thus, AI learning may have its benefits, but it also carries some downsides and risks. Whether the algorithm begets unforeseen outcomes or misapplies its learning in natural settings, an eye for adverse events and risky outcomes is required.

In the same fashion, AI's acclaimed improvements in industrial efficiency come with the risk of mass unemployment. The danger of people losing jobs to automation and robots is real. For instance, it is estimated that a single robot equals seventy human workers (Shaukat et al., 2020). Thus, while companies will benefit from faster, more reliable, and cheaper production with robots, these machines will replace human labor on a large scale. Given that robots also perform tasks with minimal error, there would be no incentive to maintain human labor. According to Shaukat et al. (2020), robots have already replaced a substantial portion of workers in blue-collar jobs and are now taking over some white-collar jobs. As AI becomes more advanced, the threat of mass unemployment will become even more salient. Importantly, current AI applications are classified as "weak AI", which not only ranks below human capabilities but also operates in restricted domains (Taeihagh, 2021, pp. 139–140). It is anticipated that this form of AI will advance into "General AI", which will be comparable to or better than human abilities and further achieve generalization in different contexts. Such powerful AI will be capable of substantially replacing human labor across sectors of the economy. Worse, AI may worsen wealth and income inequalities as its benefits are likely to be unequally distributed (Tyson & Zysman, 2022). Some countries or regions may experience a boom in employment, while others end up with mass unemployment. Without proper policies, AI's impact on employment will be devastating.

Algorithmic bias is another challenge posed by various AI technologies. Such biases risk exacerbating existing societal inequalities. For instance, some facial recognition software cannot distinguish people of color, while others can misdiagnose in healthcare settings (Alhosani & Alhashmi, 2024). The use of facial recognition software in law enforcement has been shrouded by even greater problems of bias and racial discrimination. False matches for people of color have been widely reported, reinforcing racist policing (Peter, 2021). The discrimination against racial minorities in such technologies has been attributed to biases in the training data. Problematically, the adoption of such technologies continues to grow rapidly without robust solutions. For instance, US law enforcement agencies frequently use facial recognition software to identify individuals captured in surveillance footage. According to Peter (2021), the Federal Bureau of Investigation (FBI) conducted an average of 4,000 facial recognition searches per month. The potential to worsen inequalities through algorithmic bias amid such massive usage is immense. Overall, AI bias and discrimination remain a critical challenge.

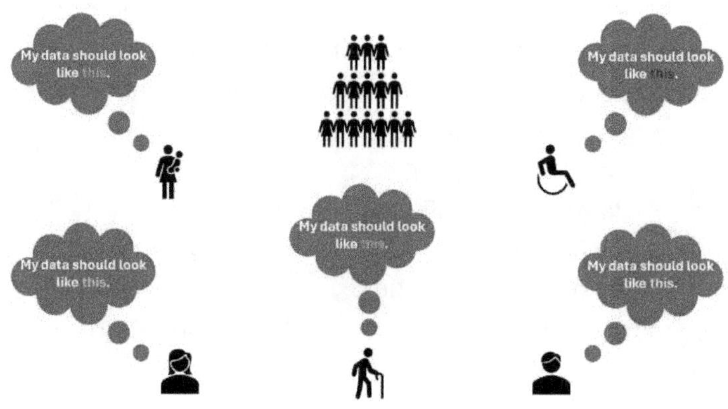

Figure 4 Everyone's Different Perception of the Data

More concerning with AI is the loss of human autonomy. This problem can take many forms. Firstly, personalization algorithms on social media and other digital platforms filter information based on user preferences, reducing the diversity of information available to such individuals (Taeihagh, 2021). This ultimately influences an individual's choices in many aspects and effectively vitiates their autonomy. This can be dangerous for society, given that important public processes like democratic elections depend on the expression of an individual's free will. The replacement of humans with AI in high-risk sectors like healthcare can also undermine human autonomy and dignity. For instance, concerns have been raised that the use of robots in providing personal care services can result in situations in which such machines restrict the subject's movements due to safety concerns (Taeihagh, 2021). Patients may end up losing their right to self-determination as AI may not be equipped to appreciate certain human values. More pressing issues have emerged in the use of autonomous weapons, such as uncrewed aerial vehicles. In particular, it is feared that such weapons can deploy lethal force against targets without moral and legal justification (Taeihagh, 2021). As such, AI may end up escalating conflicts and the use of lethal force. Consequently, the loss of human autonomy is one of the leading challenges/risks posed by AI technologies.

Loss of privacy is also a widely expressed concern over AI applications. The likelihood of leaking sensitive personal information increases as large amounts of data are used in training AI algorithms (Alhosani & Alhashmi, 2024). While steps may be taken to anonymize data, a foolproof system may be difficult to design. Equally, loss of privacy may arise from unauthorized surveillance. Most AI applications use sensors to collect data from the external environment and further rely on big data to store, analyze, and transmit through external communication networks (Taeihagh, 2021). Privacy losses may occur when AI sensors collect personal data without consent and misappropriate it. Equally, third parties may access personal data during the storage, analysis, or transmission phases and use it in unauthorized ways. This is a major concern in domains such as healthcare, where personal data stored on personal care robots and other AI applications is highly sensitive. It is also notable that AI has aided governments and corporations in increasing surveillance over their people, jeopardizing individual privacy. Companies continue to use AI applications to monitor employee performance, interfering with their basic dignity and privacy (Taeihagh, 2021). In such cases, employees may not even be free to take justified bathroom breaks out of fear of being monitored. The situation is worse with totalitarian governments that find AI to be a convenient tool for surveilling and controlling their citizens. For instance, China had about 200 million surveillance cameras in 2018 and planned to double its investment in the same (Peter, 2021). This allows the Chinese government to identify and monitor individuals in an intimate, privacy-infringing manner. With the help of AI, China was also advancing its cyber-surveillance system. AI can be used to identify dissenters on social media, notify government officials of critical messages, and circulate propaganda (Peter, 2021). Social media users and activists can no longer enjoy their privacy away from government interference. Clearly, privacy infringement is already a major problem with AI in different contexts and must be addressed going forward.

Apart from the challenges identified above, AI is also associated with risks of extinction. These are speculative scenarios that might arise as AI advances, leading to global catastrophe and the end of civilization. According to Turchin & Denkenberger (2020), these catastrophes are wide-ranging and can happen at different stages of development. Certain risks can emerge before AI begins self-improvement, some during takeoff, and others after it successfully takes over the world. It is presumed that AI will undergo the said stages of development, orchestrating the anticipated risks at each stage. For instance, a narrow AI virus could pose global catastrophic risks if it affects strategic, powerful hardware systems (Turchin & Denkenberger, 2020). This may be witnessed if such viruses target nuclear weapons, which will effectively trigger a series of events leading to mass extinction. Equally, narrow AI can lead to a large global surveillance system akin to Orwellian totalitarianism (Turchin & Denkenberger, 2020). Such serious threats are possible without self-improving AI. At advanced stages of AI development, more catastrophic outcomes are possible. For instance, Turchin & Denkenberger (2020) contend that advanced forms of AI may enslave humans through brain-infecting viruses, thereby completely eradicating their autonomy. Put simply, advanced AI systems may eventually be too powerful for humans and orchestrate their extinction.

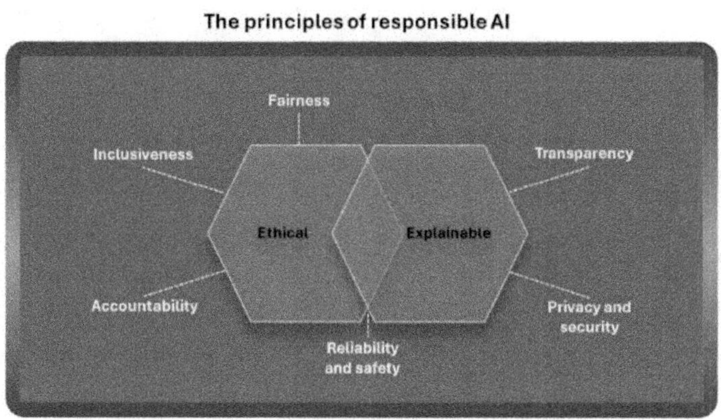

Figure 5 Principles of Responsible AI

15.3 Responsible AI Principles

Responsible AI is a governance framework that aims to maximize the benefits of AI technologies while minimizing their adverse effects. Understanding that AI is here to stay, developing guidelines for ethical/responsible AI systems is an urgent priority. Consequently, government agencies, non-governmental organizations, private companies, research institutions, and other stakeholders have published over a hundred principles of responsible AI (Hickok, 2021). While such efforts are laudable, the "principle proliferation" might create more confusion through overlaps and redundancies. Thus, it is necessary to analyze the available principles, identify common themes, and break them down into a sensible, practical number. Thus, this section unpacks some of the most salient principles of responsible AI.

One of the key principles of responsible AI is transparency. According to Eitel-Porter (2021), transparency requires that stakeholders understand how an AI model works. This principle is important because the workings of AI are often obscured even to the best experts. One might not reliably tell what data has been used, how it has been used, and how the analysis connects to the results. The "Black-Box" challenge arises when AI technologies operate on opaque evidence, leading to issues such as poor risk management and a lack of accountability. Without transparency, it is difficult to assess AI risks or assign responsibility for any harm.

Equally, a lack of transparency hinders AI adoption by different stakeholders. For instance, most consumers would be comfortable sharing their digital data if there were adequate disclosure regarding its collection and use for marketing purposes (Wang et al., 2020). This means that a lack of transparency hinders consumers from availing their data for AI applications. The principle of transparency may be expressed in different guidelines and toolkits using terms such as "intelligibility", "interpretability", "explicability", and many others. Regardless of how the principle is expressed, the main theme remains ensuring that the workings of AI are clear to all stakeholders.

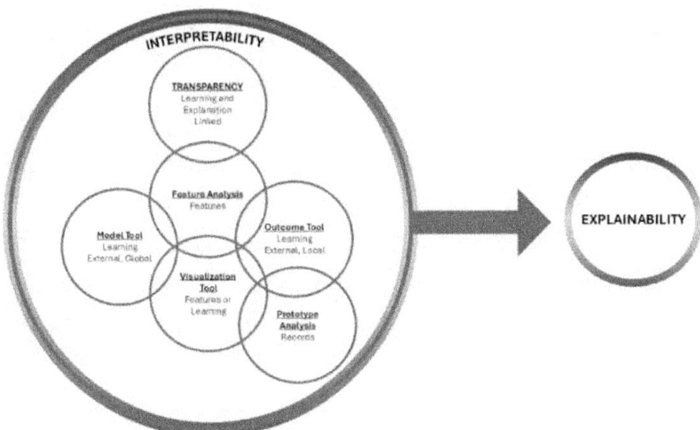

An equally important principle of responsible AI is fairness. Evidence presented by Hickok (2021) suggests that most AI ethics guidelines contain a principle of fairness and non-discrimination. In some instruments, the principle may be expressed as "justice." The ubiquity of the principle in ethics documents rests on the understanding that AI can lead to biased and discriminatory outcomes. Such cases have been widely documented in facial recognition software, AI chatbots, and other systems. Fairness requires designers of AI systems to avoid unjust outcomes, shun factors that can lead to injustice in AI models, and deliver similar results for different groups where necessary (Eitel-Porter, 2021). Fairness has many dimensions, but generally requires that AI models foster inclusion and diversity rather than promote discriminatory outcomes. According to Mikalef et al. (2022), fairness also requires consideration of how AI systems can be used to uncover human biases across different contexts. This means that AI can play a proactive role in fighting bias and discrimination rather than merely avoiding such outcomes. As observed by Tyson & Zysman (2022), AI can exacerbate existing inequalities in wealth and income, which further justifies the need to program AI to uncover such problems. If AI can detect such biases, it will help to address rather than accentuate them. Overall, responsible AI must be built on the principle of fairness.

Responsible AI must also be built on the principle of accountability. The latter concerns who takes responsibility for AI's actions (Eitel-Porter, 2021). This principle is important for various reasons. First, there is a need to apportion liability for the unjust or harmful outcomes of AI. This means that accountability necessarily connects with the principle of transparency. If the processes, rationale, and outcomes of AI are not clearly documented, it will be impossible to tell who is responsible. Secondly, accountability can be unclear when technology is used to support human decision-making (Eitel-Porter, 2021). There is a need to distinguish what human actors are responsible for and outcomes that are purely AI-generated. This also extends to AI design, as it becomes highly desirable to develop systems in which individuals, rather than algorithms, bear the most responsibility. According to Mikalef et al. (2022), accountability is crucial in domains where outcomes are sensitive, such as healthcare. For instance, when AI harms patients, it should be clear who is responsible. However, regardless of the context, accountability needs to impact both the design and deployment of AI systems.

Privacy is also important in building responsible AI. AI models rely on personal data, which is used at different stages of their life cycle. For instance, Alhosani & Alhashmi (2024) explain that the volume of data used to train AI models heightens the risk of privacy breaches. While some of the breaches may be inadvertent, measures should be taken to safeguard privacy rights in such contexts. At the operational stage, AI should be designed to safeguard rather than infringe on privacy. Eitel-Porter (2021) contends that AI models should avoid inferences that infringe on privacy. This means that such models should be designed to avoid any actions, no matter how expedient, if accomplishing them would require privacy breaches. This is necessary in domains such as healthcare, where AI may be tuned to act beneficially without regard to patient dignity and privacy. Privacy is further an important principle in light of the mass surveillance programs in countries like China (Peter, 2021). The rationale for such programs and the extent of their use must be assessed through the lens of privacy rights. The same applies to surveillance programs that organizations use to monitor employee performance. Overall, privacy is an indispensable principle in building responsible AI.

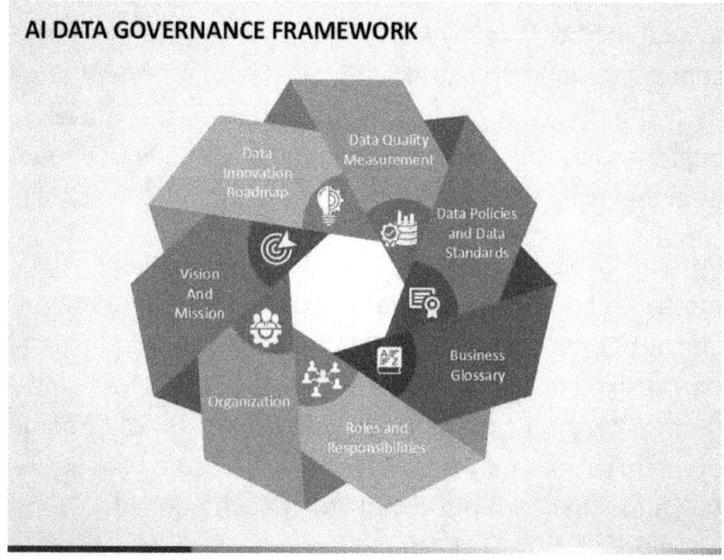

Figure 6 AI Data Framework

Most guidelines for responsible AI also include a principle of non-maleficence. Indeed, a review of the principal documents compiled by Hickok (2021) confirms that non-maleficence is a central theme permeating these guidelines. This principle requires that AI systems are not designed to cause harm or misused for nefarious ends. At its core is the recognition that AI can be remarkably harmful in the wrong hands or on its own motion. For instance, self-improving AI can develop in ways that become harmful to human beings (Turchin & Denkenberger, 2020). Designers of AI models must be conscious of such possibilities and take active steps to avoid them. Díaz-Rodríguez et al. (2023) add that AI systems should not cause or exacerbate harm to human beings and should instead protect their mental and physical integrity. The above reasoning is based on the recognition that AI can entrench inequalities and injustice, even if it does not originate them; consequently, non-maleficence overlaps with principles like fairness, accountability, and privacy. For instance, privacy infringement may be viewed as a form of harm that AI should be designed against. Some guidelines include a separate principle of robustness and/or safety. However, Díaz-Rodríguez et al. (2023) demonstrate that this falls under the umbrella of non-maleficence, requiring that AI systems be technically sound to avoid malicious misuse. On the whole, responsible AI must adhere to the principle of non-maleficence by ensuring that it does not perpetuate or exacerbate harm.

Finally, human oversight must be ensured in building responsible AI. One of the most recognized dangers of AI is its potential to violate human autonomy. Turchin & Denkenberger (2020) suggest that advanced AI models may even be capable of enslaving human beings. Even in rudimentary forms of automation, AI can take away control from humans. This necessitates building AI systems that are subject to human control. In the industrial sector, Tyson & Zysman (2022) contend that AI can be designed to complement rather than replace humans. It is possible to achieve the desired efficiency in AI systems when they are complementary to humans. This is the right course of action if human autonomy is to be preserved. However, Mikalef et al. (2022) clarify that human oversight does not mean taking away AI's autonomy. What is needed is a balance between AI autonomy and safe/responsible outcomes. Put differently, human autonomy should be preserved while maximizing the use of AI. Different forms of human oversight are necessary to ensure that the dark side of AI is kept in control.

15.4 Principles and Frameworks of AI Governance

15.4.1 The Why and How of AI Governance

Artificial Intelligence (AI) governance is a multifaceted domain that ensures AI technology is developed and operated in an ethical, transparent, and accountable manner. It comprises principles and frameworks that provide a structured approach to aligning AI systems with human values and societal norms. The OECD Guidelines, UNESCO recommendations, and Fair Information Practice

Principles (FIPs) are foundational principles that articulate the values AI should uphold, such as respect for human rights, inclusivity, transparency, and accountability. These principles are operationalized through the NIST AI Risk Management Framework, IEEE 7000-21 standard, and ISO 42001. The frameworks offer concrete steps and methodologies for implementing, identifying, and mitigating AI risks, addressing ethical concerns during system design, and providing processes to manage system standards, helping organizations establish programs to use AI responsibly. Integrating the principles and frameworks is necessary for establishing AI governance because the principles provide a set of visions and goals, while the framework offers a roadmap to achieve them. Therefore, principles and frameworks are essential in AI governance because they provide synergy that enables a holistic approach.

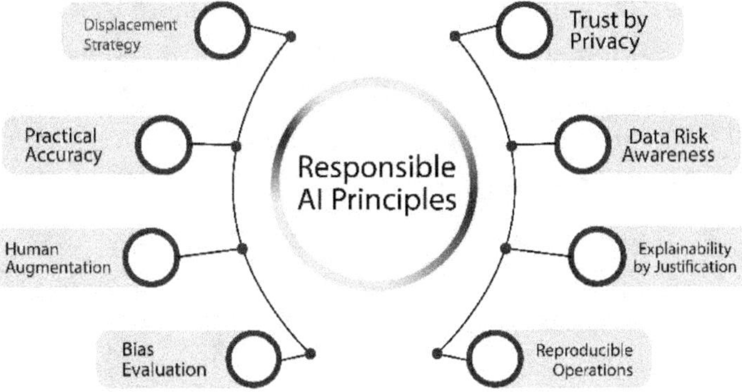

Figure 7 Responsible AI Principles

15.5 Principles of AI Governance (Whys)

15.5.1 OECD Guidelines on AI

Principles are essential in AI governance because they are the foundation for a strong structure, ensuring the ethical utilization of AI technologies. Several institutions have introduced principles to guide the development of ethical, transparent, and accountable AI technologies. For instance, the OECD guidelines on AI provide a comprehensive set of principles to ensure that AI systems are trustworthy and steer the development and deployment of AI in a manner that maximizes benefits while minimizing risks (OECD Observer, 2020). The guidelines comprise five principles that emphasize that AI should be inclusive and sustainable, contributing positively to societal growth and environmental stewardship. They also advocate protecting human rights and implementing fair practices to ensure that AI respects the inherent dignity and equality of all individuals. The importance of transparent and explainable AI is also highlighted to allow individuals to understand and trust the AI systems they use.

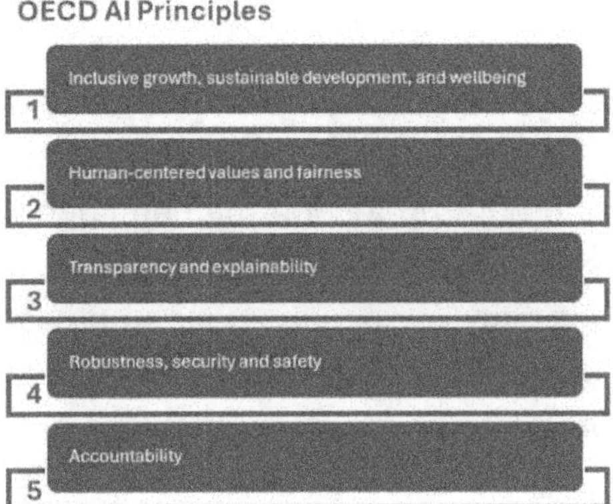

Figure 8 OECD AI Principles

Further, the guidelines call for robustness, security, and safety in AI systems to ensure they are reliable and protected against threats. Finally, they emphasize that developers should be accountable for the systems they are introducing; hence, they provide a mechanism for recourse and redress where necessary. These policies serve as a blueprint for responsible AI that aligns with human values and societal goals.

15.5.2 UNESCO's Recommendation on Ethics of AI

Adopted by UNESCO in November 2021, the UNESCO Recommendations were crafted by the Ad Hoc Expert Group (AHEG). This document outlines ten recommendations to ensure the ethical and safe use of AI, emphasizing its impacts on culture, education, science, information, and communication. The organization recommends that AI emphasizes proportionality, safety and security, equity and non-discrimination, sustainability, right to privacy and data protection, human supervision, transparency and explainability, awareness and education, responsibility and accountability, and adaptive governance and collaboration. The principles are designed to ensure that AI systems prioritize human rights and ethical considerations, aligning with the fundamental values of respect and fairness. There is also a strong focus on supporting human autonomy and decision-making, reflecting the importance of AI as a tool to improve human capabilities rather than replace them. UNESCO's strong emphasis on data privacy is prioritized because of the risk of personal data abuse. Moreover, promoting equity and avoiding biases are necessary to prevent the perpetuation of historical injustices and to ensure that AI benefits all segments of society. Finally, ensuring that AI systems are free from discrimination contributes to equitable service for the global population. These recommendations reflect a commitment to harnessing the potential of AI while safeguarding against its risks, ensuring that it serves the common good.

FAIR INFORMATION PRACTICE PRINCIPLES

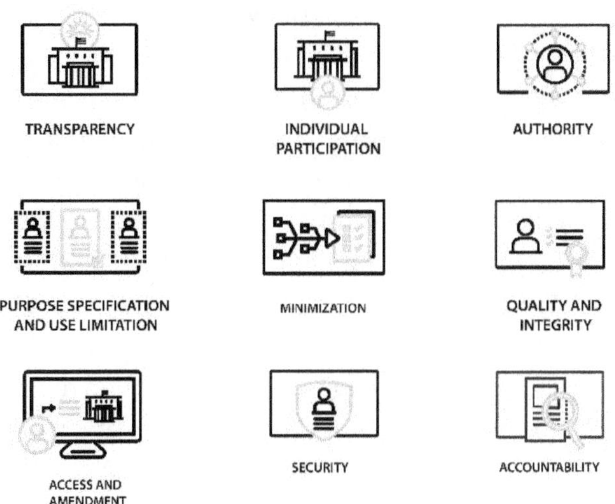

Figure 9 Fair Information Practice Principles

15.5.3 Fair Information Practices (FIPs)

The FIPs principles are at the core of the Privacy Act of 1974 and have been adopted by most U.S. states. In AI development, the principles set guidelines to protect personal data and ensure privacy in the digital age (Harper, 2021). FIPs comprise eight principles, including collection limitation, data quality, purpose specification, use limitation, security safeguards, openness, individual participation, and accountability. They are the foundation for data governance frameworks worldwide because they emphasize the importance of ethically and responsibly managing data. The principle of collection limitation prevents overreaching by ensuring that only essential data is collected. Data quality ensures that the information collected is accurate and relevant, enabling informed decisions. Purpose specification demands that the reasons for collecting data are explicit and legitimate to prevent misuse. Use limitations restrict the application of data to the purpose specified at the time of collection, safeguarding against unauthorized use. Security safeguards protect data from threats to ensure confidentiality, integrity, and availability. Openness about data practices fosters trust and allows for informed consent (Harper, 2021). Individual participation recognizes the right of the individual to access their data and make corrections if necessary. Finally, accountability ensures that those who control data are answerable for adhering to these principles, providing a mechanism for enforcement and redress. Therefore, together, the principles form a robust framework for data governance, which balances the need of organizations to collect and use data with the rights of individuals to privacy and control over personal information.

15.6 Frameworks for AI Governance (Hows)

15.6.1   NIST   AI   Risk   Management Framework

The AI governance framework helps operationalize AI governance principles to ensure organizations comply with established principles and regulations. Several frameworks have been introduced to help organizations in this endeavor. For example, the AI Risk Management Framework (RFM) was developed by the National Institute of Standards and Technology (NIST). AI RFM provides a systematic approach to risk management to help organizations comply with regulatory requirements and foster trust and confidence in AI systems among stakeholders (U.S. Department of Commerce & NIST, 2023). The framework emphasizes mapping, measuring, managing, and governing risks to ensure that AI systems are safe, secure, and aligned with ethical standards. Although the method's effectiveness has not been evaluated, it is speculated that the framework will benefit users in numerous areas. As AI continues to evolve, frameworks such as AI RFM will help organizations navigate the complex landscape of AI governance, thereby benefiting society while minimizing potential harm. AI governance is an ongoing process that requires vigilance, adaptability, and a commitment to continuous improvement. Therefore, by focusing on framing and addressing risks and potential impacts on users along the AI lifecycle, this framework aims to enhance the trustworthiness of AI systems.

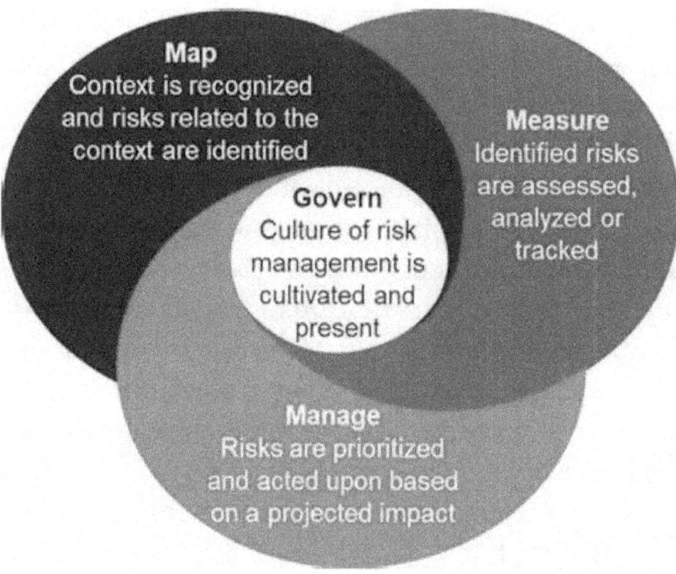

Figure 10 NIST Management Framework

15.6.2   IEEE 7000-21 Process for Addressing Ethical Concerns

The IEEE 7000-21 standard establishes ideals that organizations can use when considering integrating ethical values into the concept exploration and development stages. The framework supports transparent communication with selected management and engineering stakeholders to elicit and prioritize ethical values. The standard highlights the need to consider technology's wider impact on society and the duty of engineers to tackle these issues proactively. Notably, by providing engineers and technologists with implementable processes and aligning them with innovation management process systems, approaches, and software engineering methods, the standard helps address concerns and risks during system design. Moreover, by engaging stakeholders, establishing core values, evaluating ethical impacts, creating mitigation strategies, and continuously monitoring results, engineers can ensure their designs meet technical requirements and adhere to social and ethical standards. The process fosters a culture of responsibility and accountability in technological development, which is vital in an era where technology increasingly shapes every aspect of life.

IEEE Standard Model Processes for Addressing Ethical Concerns During System Design

Concept exploration stage		Development stage	
Concept of operations and context exploration processes	Ethical values elicitation and prioritization process	Ethical requirements definition process	Ethical risk-based design processes
Transparency management process			

Figure 11 IEEE Standard Model

15.6.3 ISO 42001 (Management Systems)

ISO 42001 is a management system designed to guide organizations in the effective governance of AI. It is built around the Plan-Do-Check-Act (PDCA) cycle, a known iterative process used for continuous improvement (ISO, n.d.). The PDCA cycles consist of the planning, implementation, monitoring, and action phases, which collectively help organizations to ensure that their AI systems are effectively implemented, maintained, and continuously improved. Since it proposes a single model, the framework allows the standardization of AI systems to ensure they comply with governance principles. The framework consists of several key elements, including policy development, risk management, performance evaluation, nonstop feedback, and continuous evaluation (ISO, n.d.). Policy development establishes clear policies and sets a foundation for ethical AI use, reflecting the core governance principles. The risk management element helps identify potential issues and implement strategies to mitigate them, thus safeguarding against misuse and ensuring reliability. Performance evaluation is another cornerstone of the ISO 42001 framework, providing metrics to assess the effectiveness of AI governance practices (ISO, n.d.). The continuous feedback loop helps refine AI systems and ensure they remain effective and ethical. Lastly, the principle of continuous improvement emphasizes the dynamic nature of AI and the need for policies and practices to evolve with technological advancements and societal expectations, which maximizes AI's benefits while minimizing its risks. This standard is a blueprint for responsible AI deployment, balancing innovation with accountability.

ISO 42001 Benefits

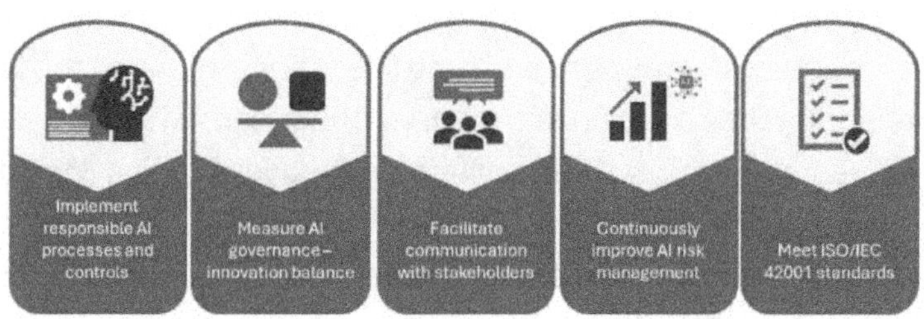

Figure 12 ISO 42001 Benefits

15.6.4   HUDERAF (Human Rights, Democracy, Laws, and Assurance)

The HUDERAF framework is a practical toolkit designed to integrate human rights into AI governance to ensure trustworthy AI innovation practices. The framework provides a comprehensive end-to-end approach to maintaining the integrity of AI project lifecycles. It encompasses context-specific risk analysis, stakeholder involvement, thorough impact evaluation, clear risk management, impact reduction, and assurance of innovation. Essentially, the human rights element of the framework ensures that AI systems are designed and operated in a manner that respects individuals' inherent dignity and rights. Democratic oversight is crucial for maintaining public trust and allowing for transparency and accountability in AI systems. Adhering to legal regulations is crucial, as it ensures that AI activities align with current laws and rules, serving as a primary safeguard against potential misuse. Lastly, ethical assurance is about embedding moral principles into the fabric of AI systems, ensuring they operate in a manner consistent with societal values and norms. These pillars collectively form a robust structure for the ethical and lawful use of AI, reflecting a commitment to upholding the rule of law and democratic principles in the digital age.

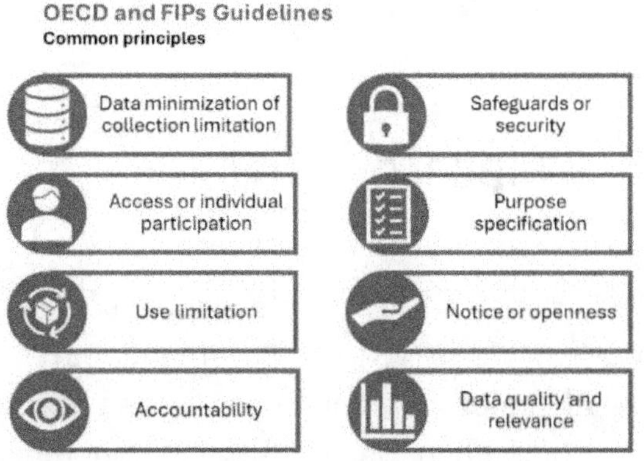

Figure 13 OECD and FIPS

15.7 The Interplay Between Principles and Frameworks

Principles and frameworks work together to ensure that AI develops ethically and securely and benefits society. Principles set the ethical direction for AI, while frameworks translate these ideals into actionable steps and structure their implementation. For instance, the AI RMF integrates international guidelines into national standards, supporting robustness, security, and accountability. Frameworks like IEEE 7000-21 reflect UNESCO's human-centered design principles, promoting fairness and inclusivity in AI. Together, principles provide guidance and frameworks operationalize them, both vital for effective AI governance.

The ISO 42001 framework integrates principles and management systems to promote transparency, accountability, and data quality in AI governance. It features policy development, risk management, performance evaluation, and continuous improvement, reflecting core FIPs of data protection (Mittelstadt, 2019). For instance, ISO 42001 requires organizations to clearly communicate AI data practices, establish accountability, and ensure compliance with ethical standards. By standardizing data governance, it supports accurate, secure use of data. UNESCO principles and the HUDERAF framework similarly emphasize human rights, democratic oversight, legal compliance, and ethical values, highlighting that effective AI governance goes beyond compliance to build a culture of responsibility and ongoing improvement.

Figure 14 Synergy between Values and Recommendations

Importance of the Synergy Between Principles and Framework

Integrating principles and frameworks is key to effective AI governance. By putting principles into practice, AI systems can align with human rights and societal values and earn public trust. Governance frameworks minimize risks, address biases and security issues, and help ensure legal compliance. Strong governance also supports innovation and inclusivity, aiming to distribute AI's benefits widely and tackle global challenges. Ultimately, AI governance upholds ethical use of technology for a positive future.

15.8 Conclusion

AI governance, encompassing principles and frameworks, plays a crucial role in ensuring that AI systems are developed for the benefit of society. Principles serve as an ethical foundation, while frameworks provide practical methodologies for implementing these values. Guidelines from organizations such as the OECD and UNESCO, along with FIPs, outline standards that AI systems should adhere to in order to remain ethical and beneficial. Conversely, frameworks such as NIST AI RMF, ISO 42001, and HUDERIA offer structured approaches to operationalizing the values articulated by these principles. Notably, principles and frameworks work together: principles establish objectives, and frameworks supply the roadmap for achieving them. Analysis indicates that the synergy between principles and frameworks yields robust AI governance practices, fostering ethical development in AI systems. Effective alignment of these elements contributes to the creation of strong, secure, and accountable systems, collectively referred to as AI governance. Ultimately, AI governance is designed to mitigate risks and advance an environment in which AI technology can reliably and constructively evolve.

16 The Legal Framework

This section will provide an overview of the nature of the US legislative and regulatory regime for AI, delineate existing laws, and further outline emerging legislation on different aspects of AI.

16.1 Overview of the US Approach

AI regulation differs globally. The US generally adopts soft law and private-sector self-regulation, favoring free-market principles over strict legislation, while the EU prefers hard law (Castets-Renard, 2021). Due to the US's federal structure, states have primary legal authority in many areas, resulting in varied and sometimes conflicting regulations (Yoo & Lai, 2020). Soft laws help bridge these jurisdictional gaps. Most US proposals focus on specific AI sectors, such as autonomous vehicles or facial recognition, with limited attention to broader AI issues, such as training data. However, substantial legal precedents exist for AI in the US. Below is a summary of current and emerging legislation.

16.2 Existing Laws

Laws and regulations governing autonomous vehicles are notable elements of the legal framework for AI in the US. However, most regulations and laws are state-level, ostensibly because transportation falls under state, rather than federal, powers (Castets-Renard, 2021). The federal government only intervenes in the realm of interstate transportation and the setting of safety measures, with the latter powers already exercised under the Motor Vehicle Safety Act of 1966. Consequently, the states have taken a proactive role in regulating autonomous vehicles, leading to a complex patchwork of legislation. According to Mallinson et al. (2024), there are discernible regional patterns in states' approaches to regulation. The authors explain that Northeastern states tread cautiously, unlike their southern counterparts, such as Texas, which have more permissive laws regarding the testing and deployment of autonomous vehicles. On their part, California and Washington have the most progressive laws on the subject, positioning them as clear leaders in the development of autonomous vehicles (Mallinson et al., 2024). Available laws in different states cover issues like vehicle data privacy, testing, licensing and registration, and insurance and liability. Most legislation encourages the deployment of autonomous vehicles, though Castets-Renard (2021) warns that there is insufficient focus on privacy and safety. The regulation of autonomous vehicles in the US is overall substantial, though it evidently needs refinement.

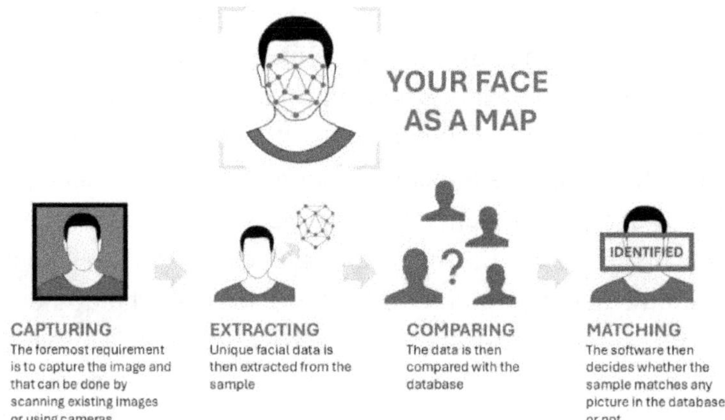

Figure 15 Facial Recognition

Facial recognition software is another aspect of AI that has been the subject of several legislative and regulatory actions. Just like autonomous vehicles, the technology has so far been controlled by states and local authorities rather than the federal government. Unfortunately, most of the regulatory actions have taken the form of outright bans. San Francisco was the first city to ban the use of facial recognition software by public authorities in 2019, a move that was replicated by Boston, Northampton, Oakland, Springfield, and many others (Castets-Renard, 2021). These bans were introduced through ordinances passed by city councilors that outlawed the use of the technology in police and other city departments, citing negative effects such as misidentification and endangering public safety. States have taken a similar turn, proposing a moratorium on facial recognition software or banning its use in certain contexts. For instance, California and Oregon have passed laws prohibiting law enforcement officers from using facial recognition software in their body cameras (Castets-Renard, 2021). Other states have legislative proposals that are aligned with this. There is an urgent need for laws that promote the ethical use of such technologies, rather than outright bans.

Various federal agencies have statutory powers that can lead to significant AI regulation. These agencies, empowered by underlying laws, can create rules to address new AI scenarios. For example, the FDA oversees AI tools in healthcare as it regulates medical devices under the 21st Century Cures Act of 2016. AI-powered medical devices must thus obtain FDA approval before being launched into the market. This may hinder AI adoption as the FDA may be reluctant to approve devices that are not fully explicable or intended to operate without human supervision. However, Yoo & Lai (2020) observe that the FDA has rolled out a new regulatory framework to address these developments in AI. The FDA's rulemaking powers ensure it can continuously improve its testing and approval procedures to promote meaningful developments in AI. Other federal agencies, such as the Department of Transportation (DOT), have issued guidelines within their statutory mandates that states may voluntarily adopt to harmonize laws on autonomous vehicles (Yoo & Lai, 2020). The DOT can also go a step further and develop rules, as long as it does not trespass on the states' powers. Thus, some AI regulation may flow from existing statutory powers devolved to various federal agencies.

A review of US laws on AI must consider how constitutional provisions impact the technology. These constitutional rights, such as privacy and non-discrimination, act as direct regulators of AI, allowing individuals to seek legal remedies if AI violates their guaranteed freedoms. For example, privacy breaches or unfair criminal attribution by AI can lead to legal action. Although AI's complexity may pose challenges for direct constitutional application, the Constitution still influences both AI regulation and legislation. Strong First Amendment protections limit algorithmic content moderation, and platforms compelled to use AI for this purpose could face constitutional challenges. Ultimately, the Constitution sets clear standards relevant to AI governance.

16.3 Emerging Laws

Federal inaction on AI legislation risks a patchwork of contradictory state laws, which may impede both innovation and enforcement. Fortunately, several legislative proposals addressing different aspects of AI are currently in both the House and the Senate. However, this does not mean that the states and local authorities are not enforcing key AI laws. A review of upcoming laws for governing AI at all levels of government is as follows.

16.3.1.1 Federal laws

Key among the emerging federal-level laws are legislative proposals to promote AI research. Notably, the Growing Artificial Intelligence Through Research Act (GrAITR) was introduced in the House back in 2019 (Yoo & Lai, 2020). The proposed legislation enjoyed bipartisan support and sought to direct the President to spearhead a national initiative to invest in AI, increase the volume of skilled workers in the area, and encourage data sharing between state and non-state agencies. The Act also contemplates a national education program on AI and engineering and provides grants to establish multidisciplinary research centers (Yoo & Lai, 2020). Legislation of this kind at the federal level is heartening as it signposts the government's intention to commit its resources to advancing the understanding and beneficial application of AI. The Artificial Intelligence Initiative Act (AI-IA), with similar goals, was subsequently introduced in the Senate, and its proposals differed only slightly from those of GrAITR (Yoo & Lai, 2020). The two laws represent the most progressive proposals to advance AI research at the national level and may have a highly transformative impact if successfully enacted. However, both remain in the pipeline with no clear indication of when they shall become law.

Some federal legislation for regulating autonomous vehicles and facial recognition software has also been proposed. Unfortunately, the SELF-DRIVE Act, which was bound to become the first federal law to regulate autonomous vehicles, died in the Senate (Green, 2020).

The Act would have provided a national, unifying character. The House Energy and Commerce Committee unanimously passed the bill. Still, upon referral to the Senate's Committee on Commerce, Science, and Transportation, it was never brought to the full Senate for a vote (Green, 2020). It is unclear why the Senate missed the opportunity to pass such an important law. However, there is hope for legislation meant to regulate facial recognition software. Crucially, the Commercial Facial Recognition Privacy Act was introduced in the Senate in 2019, with the bill intended to prohibit unauthorized use of facial recognition technology by private entities (Castets-Renard, 2021). The Act would also outlaw sharing facial recognition data with third parties without the subject's consent. Later in 2020, the Ethical Use of Facial Recognition Act was introduced into the Senate, a law that would prohibit the use of facial recognition by a federal officer, employee, or contractor without a warrant until rules governing its public or commercial use were promulgated (Castets-Renard, 2021). It is encouraging that the proposed laws were not entirely prohibitive of the use of the said technologies and, to some extent, encouraged their responsible use. However, until they are passed, there is a need to remain guardedly optimistic.

Also proposed are laws to address the privacy threat posed by AI systems. Key among them is the Consumer Online Privacy Rights Act (COPRA), which requires organizations to conduct impact assessments when deploying algorithms to determine eligibility for services like housing and credit (Yoo & Lai, 2020). COPRA would also mandate such assessments for all places of public accommodation. The bill, however, has not yet progressed beyond the committee stage. Another important proposal is the Data Protection Act (DPA), which was first introduced into the Senate by Kirsten Gillibrand in 2020 (Cook, 2022). The bill would have established an independent federal agency to enforce rules for the protection of individual data. The agency would also take complaints and carry out investigations on matters of data protection. However, the bill never made it beyond the committee stage and died with the 116th Congress (Cook, 2022). Undeterred, Kirsten Gillibrand reintroduced the bill in 2021. The new proposal is said to include significant improvements, including protection against high-risk data practices and remedies to privacy harms and discrimination (Cook, 2022). The new version thus demonstrates a greater awareness of AI risks to privacy. However, it remains to be seen if the 117th Congress will pass the bill. Overall, the proposed federal laws indicate a positive step towards regulating the privacy risks posed by AI.

Figure 16 Algorithmic Accountability Act

Equally important in regulating AI is the Algorithmic Accountability Act of 2022. The bill is a revision of the 2019 version and focuses on regulating automated decision systems (ADS) used by organizations (Mökander, Juneja, et al., 2022). Specifically, the proposed law requires organizations using ADS to take proactive steps to identify and mitigate social, ethical, and legal risks inherent in such technologies. The bill represents a significant federal effort to regulate automation across different industries and appears to depart from the erstwhile insistence on self-regulation (Mökander, Juneja, et al., 2022). Importantly, the proposed law includes comprehensive provisions that could help address some of the main problems associated with AI systems to date. For instance, the law requires organizations to fix algorithms that produce unfair, discriminatory, or inaccurate outcomes (Yoo & Lai, 2020). This reflects a deep awareness of the ethical issues AI has raised across different sectors. The provision is also indicative of legislative prescriptions for AI design, which have been absent from past laws. The Algorithmic Accountability Act has been lauded for providing mechanisms to hold bad actors accountable while enabling those with good intentions to establish safe, ethical, and legally sound systems (Mökander, Juneja, et al., 2022). This is a welcome departure from precautionary approaches that impose absolute bans on such technologies. If the bill successfully sails through the Senate, it is likely to be a landmark piece of legislation that will set the pace for AI regulation globally.

16.3.1.2 State Initiatives

It is also important to note that states are passing AI legislation to regulate various aspects of the technology. The approach to legislation across the states is varied, just like Mallinson et al. (2024) observed in the case of autonomous vehicles. Each state is adopting a regulatory approach that best aligns with its assessment of regulatory needs. For instance, California and New Jersey have passed laws that are aimed at mitigating the risks posed by AI decision-making (Shen, 2023). The laws appear coterminous with the Algorithmic Accountability Act at the federal level but focus on physical infrastructure. Washington has proposed a more robust ADS law, which Shen (2023) considers a pioneer for AI regulation in the public sector. The proposed law prohibits the development or use of ADS to discriminate against individuals. Further, it requires completion of an accountability report prior to the procurement, development, or use of ADS. Some states are enacting mild legislation that does not directly address the impacts of AI. A good example is New York, whose proposed AI law merely creates a commission to research how to regulate AI (Shen, 2023). States like New York do not appear to appreciate the urgency of regulating AI. Overall, there is an expectation that a large volume of disparate AI laws will emerge from the states in the foreseeable future.

16.3.2   Chapter Conclusion

Ultimately, it appears that the US will firmly embrace the hard approach to regulating AI and move away from self-regulation and soft law. A flurry of legislation at the federal and state levels may result in many laws regulating various aspects of AI. However, this might lead to overregulation, as Castets-Renard (2021) warned, thereby stifling innovation. A complex patchwork of state and federal regulations will only complicate compliance and increase administrative costs. On the flip side, many of the laws may not see the light of day, maintaining the equally undesirable under-regulated environment. Importantly, Cook (2022) demonstrates that key bills, such as the Data Protection Act, have died in the Senate, reflecting legislative reticence to regulate AI. The same problem is evident in the states as well. For instance, Shen (2023) observed that only four out of seventeen states that had proposed AI/ADS legislation had enacted it. This suggests that many states were still reluctant to take bold steps in regulating AI. Both overregulation and underregulating are undesirable, and efforts should be made to ensure that a balanced level of regulation is achieved in the end.

17 Content References

Ada Lovelace Institute. *Regulating AI in Practice: Lessons from Early Implementers*. Ada Lovelace Institute, 2024. (CH1, CH7)

Algorithmic Justice League. *Auditing AI Systems for Bias at Scale*. Algorithmic Justice League, 2023. (CH4)

Almeida, Fernando, and Sofia Silva. "AI Governance Models for the Public Sector." *Government Information Quarterly*, vol. 41, no. 2, 2024. (CH1, CH9)

American Bar Association. *AI, Liability, and Emerging Legal Duties*. American Bar Association, 2024. (CH6, CH16)

Australian Human Rights Commission. *Human Rights and Artificial Intelligence: Updated Guidance for 2024*. Australian Human Rights Commission, 2024. (CH3, CH4, CH15)

Binns, Reuben. "Fairness in Machine Learning: Recent Developments and Practical Challenges." *AI and Ethics*, vol. 4, 2024. (CH4)

Borgesius, Frederik Zuiderveen. "Strengthening Transparency in Algorithmic Decision-Making." *International Data Privacy Law*, vol. 13, no. 3, 2023. (CH5)

Brundage, Miles, et al. "Managing AI Risks in Practice: A Systems Approach." *Harvard Kennedy School M-RCBG Working Paper*, 2023. (CH3, CH8)

Bryson, Joanna J. *The Governance of Artificial Intelligence: Ethics, Accountability, and Power*. Oxford UP, 2023. (CH1, CH6, CH15)

Calo, Ryan, and Danielle Citron. "The Law of Artificial Intelligence in 2025." *California Law Review*, vol. 113, 2025. (CH7, CH16)

Cath, Corinne. "AI, Power, and Democratic Oversight." *Internet Policy Review*, vol. 12, no. 1, 2023. (CH6, CH15)

Centre for Data Ethics and Innovation. *Responsible AI in the Public Sector: Implementation Guide 2023–2025*. UK Dept. for Science, Innovation and Technology, 2023. (CH1, CH9)

Cihon, Peter, et al. "Standards for AI Governance: A Global Mapping." *Journal of Responsible Technology*, vol. 15, 2024. (CH8, CH15)

Crawford, Kate. *Atlas of AI: Power, Politics, and the Planetary Costs of Artificial Intelligence. Updated Ed.* Yale UP, 2024. (CH3, CH11, CH15)

Data & Trust Alliance. *Responsible AI Playbook for Enterprises: Second Edition*. Data & Trust Alliance, 2024. (CH1, CH8, CH9)

Dignum, Virginia. *Responsible Artificial Intelligence in Practice: Governance and Operations*. Springer, 2023. (CH1, CH8, CH10)

European Commission. *Regulation (EU) 2024/... Artificial Intelligence Act (EU AI Act)*. Official Journal of the European Union, 2024. (CH7, CH8, CH15, CH16)

European Data Protection Board. *Guidelines on the Use of Personal Data in AI Systems*. EDPB, 2023. (CH7, CH11, CH16)

European Union Agency for Fundamental Rights. *Fundamental Rights and AI Systems: Updated Risk Perspectives*. FRA, 2023. (CH3, CH4, CH15)

Floridi, Luciano, and Josh Cowls. *The Ethics of AI: A New Framework for Governance*. Polity, 2023. (CH2, CH10, CH15)

Future of Privacy Forum. *AI and Data Protection in Practice: Risk and Governance Toolkit*. Future of Privacy Forum, 2023. (CH3, CH7, CH11)

Gasser, Urs, and Virgilio A. F. Almeida. "A Layered Model for AI Governance." *Journal of Cyber Policy*, vol. 8, no. 2, 2023. (CH9, CH15)

Harvard Business Review. *HBR's 10 Must Reads on Responsible AI*. Harvard Business Review Press, 2024. (CH1, CH2, CH3)

IEEE. *IEEE 7000-2024: Model Process for Addressing Ethical Concerns During System Design*. IEEE Standards Association, 2024. (CH2, CH10, CH15)

IEEE. *IEEE 7010-2023: Recommended Practice for Assessing the Impact of AI on Human Well-Being*. IEEE Standards Association, 2023. (CH3, CH13)

International Association of Privacy Professionals. *AI Governance and Privacy: Operational Guide for 2024*. IAPP, 2024. (CH7, CH11, CH16)

International Organization for Standardization. *ISO/IEC 23894:2023 Information Technology—Artificial Intelligence—Risk Management*. ISO, 2023. (CH3, CH8)

International Organization for Standardization. *ISO/IEC 42001:2023 Artificial Intelligence— Management System.* ISO, 2023. (CH1, CH8, CH9)

Kroll, Joshua A., et al. "Accountability in Algorithmic Systems: Updated Perspectives." *Communications of the ACM*, vol. 67, no. 5, 2024. (CH5, CH6)

Leslie, David. *Operationalizing AI Ethics: Translating Principles into Practice.* The Alan Turing Institute, 2023. (CH2, CH10, CH14)

Mittelstadt, Brent. "AI Ethics Beyond Principles: From Compliance to Practice." *AI and Society*, vol. 39, 2024. (CH2, CH10, CH12)

National Institute of Standards and Technology. *Artificial Intelligence Risk Management Framework (AI RMF 1.0).* NIST, 2023. (CH3, CH8, CH9)

National Institute of Standards and Technology. *NIST Generative AI Profile: Draft for Public Comment.* NIST, 2024. (CH3, CH8)

OECD. *OECD Framework for the Classification of AI Systems: 2024 Update.* OECD Publishing, 2024. (CH8, CH15)

OECD. *OECD Recommendation on Artificial Intelligence: Implementation and Monitoring Report 2023–2024.* OECD Publishing, 2024. (CH2, CH15)

Partnership on AI. *Responsible Practices for Synthetic Media and Generative AI*. Partnership on AI, 2023. (CH3, CH10)

Partnership on AI. *Managing AI Risks: A Framework for Organizations*. Partnership on AI, 2024. (CH3, CH8)

Raji, Inioluwa Deborah, et al. "Auditing AI Systems: A Practical Field Guide." *ACM Conference on Fairness, Accountability, and Transparency (FAccT) 2023 Proceedings*, ACM, 2023. (CH4, CH8, CH13)

Reisman, Dillon, et al. *Algorithmic Impact Assessments: A Practical Guide for Public Agencies. 2024 Update*. AI Now Institute, 2024. (CH3, CH8, CH9)

Royal Society. *From Principles to Practice: Responsible AI in the Public Sector*. The Royal Society, 2023. (CH1, CH9, CH14)

Smuha, Nathalie A. "The EU AI Act: A New Model for Global AI Regulation." *Computer Law & Security Review*, vol. 52, 2024. (CH7, CH15, CH16)

Stanford Institute for Human-Centered Artificial Intelligence. *AI Index Report 2024*. Stanford HAI, 2024. (CH3, CH13)

Tutt, Andrew, and Woodrow Barfield. *AI, Law, and Policy: Emerging Regulatory Models*. Edward Elgar, 2025. (CH7, CH15, CH16)

UNESCO. *Recommendation on the Ethics of Artificial Intelligence: Implementation Guide 2023–2025*. UNESCO, 2023. (CH2, CH12, CH15)

United Nations High Commissioner for Human Rights. *The Right to Privacy in the Digital Age: AI and Surveillance*. United Nations, 2023. (CH3, CH7, CH11)

United Nations Office of the High Commissioner for Human Rights. *Human Rights Impact of AI Systems: Updated Guidance 2024*. United Nations, 2024. (CH3, CH4, CH15)

United States, Executive Office of the President. *Executive Order on Safe, Secure, and Trustworthy Artificial Intelligence*. The White House, 2023. (CH7, CH8, CH16)

United States Federal Trade Commission. *Protecting Consumers from AI-Driven Harm: Enforcement Priorities 2024*. FTC, 2024. (CH3, CH6, CH16)

United States Government Accountability Office. *Artificial Intelligence: Key Practices to Help Ensure Accountability and Transparency*. GAO, 2023. (CH5, CH6, CH9)

World Economic Forum. *Global AI Governance Alliance: Policy and Practice Report 2024*. World Economic Forum, 2024. (CH1, CH8, CH15)

World Economic Forum. *Responsible AI Playbook for Chief Risk Officers*. World Economic Forum, 2023. (CH3, CH8)

World Health Organization. *Ethics and Governance of AI for Health: 2023–2025 Implementation Report*. WHO, 2023. (CH3, CH8, CH11)

World Bank. *AI in Public Sector Modernization: Governance, Risk, and Accountability*. World Bank, 2024. (CH1, CH8, CH9)

World Bank. *Data Governance for AI-Enabled Services in Government*. World Bank, 2023. (CH1, CH11)

World Privacy Forum. *AI, Data Brokers, and Risk to Individuals: Policy and Governance Recommendations*. World Privacy Forum, 2023. (CH3, CH7, CH11)

World Wide Web Consortium (W3C). *Technical Architecture and Governance Considerations for AI on the Web*. W3C, 2024. (CH5, CH9)

Zerilli, John, et al. "Human Oversight in AI Systems: Designing Effective Safeguards." *AI and Law*, vol. 32, 2024. (CH3, CH6, CH12)